Lecture Notes in Statistics 87

Edited by S. Fienberg, J. Gani, K. Krickeberg,
I. Olkin, and N. Wermuth

Jesper Møller

Lectures On Random Voronoi Tessellations

Springer-Verlag
New York Berlin Heidelberg London Paris
Tokyo Hong Kong Barcelona Budapest

Jesper Møller
Department of Theoretical Statistics
Institute of Mathematics
University of Aarhus
DK-8000 Aarhus C
DENMARK

Library of Congress Cataloging-in-Publication Data
Møller, Jesper.
   Lectures on random Voronoi tessellations / Jesper Møller.
      p.   cm. -- (Lecture notes in statistics ; 87)
   Includes bibliographical references and index.

   1. Voronoi polygons. 2. Spatial analysis (Statistics). I. Title.
   II. Series: Lecture notes in statistics (Springer-Verlag) ; v. 87.
   QA278.2.M64 1994
   133.9'01'3 -- dc20   94-248

Printed on acid-free paper.

Camera ready copy provided by the author.

9 8 7 6 5 4 3 2 1

ISBN-13: 978-0-387-94264-3   e-ISBN-13: 978-1-4612-2652-9
DOI: 10.1007/978-1-4612-2652-9

Editorial Policy
for the publication of monographs

In what follows all references to monographs are applicable also to multiauthorship volumes such as seminar notes.

§ 1. Lecture Notes aim to report new developments - quickly, informally, and at a high level. Monograph manuscripts should be reasonably self-contained and rounded off. Thus they may, and often will, present not only results of the author but also related work by other people. Furthermore, the manuscripts should provide sufficient motivation, examples, and applications. This clearly distinguishes Lecture Notes manuscripts from journal articles which normally are very concise. Articles intended for a journal but too long to be accepted by most journals, usually do not have this "lecture notes" character. For similar reasons it is unusual for Ph.D. theses to be accepted for the Lecture Notes series.

§ 2. Manuscripts or plans for Lecture Notes volumes should be submitted (preferably in duplicate) either to one of the series editors or to Springer-Verlag, New York. These proposals are then refereed. A final decision concerning publication can only be made on the basis of the complete manuscript, but a preliminary decision can often be based on partial information: a fairly detailed outline describing the planned contents of each chapter, and an indication of the estimated length, a bibliography, and one or two sample chapters - or a first draft of the manuscript. The editors will try to make the preliminary decision as definite as they can on the basis of the available information.

§ 3. Final manuscripts should be in English. They should contain at least 100 pages of scientific text and should include
- a table of contents;
- an informative introduction, perhaps with some historical remarks: it should be accessible to a reader not particularly familiar with the topic treated;
- a subject index: as a rule this is genuinely helpful for the reader.

Preface

These notes were prepared for a series of lectures presented at a summer course 1992 on 'Random Voronoi Tessellations' organized by Professor Julian Besag, Department of
Statistics, University of Washington.

Hopefully the notes may provide a helpful supplement to
the existing mathematical literature on random Voronoi tessellations, which might be difficult for non-specialists to enter.
The notes were prepared just before the excellent book by Okabe
et al. (1992) appeared. While Okabe et al. (1992) contains a
comprehensive collection of concepts and results of Voronoi
tessellations but often without mathematical details and
proofs, the exposition in the present booklet is mathematically
rigorous and detailed proofs are given making the notes largely
self-contained. Also no background knowledge on either the
subject of this booklet or of spatial statistics and stochastic
geometry is assumed.

Finally, I am indebted to Julian Besag for giving me the
opportunity to visit Seattle and Department of Statistics,
University of Washington, and to Oddbjørg Wethelund for her
eminent secretarial assistance.

December 1992

Jesper Møller

Contents   Page

Preface

1. Introduction and background

A tessellation or mosaic of the d-dimensional Euclidean space $\mathbb{R}^d$ is a subdivision $\mathbb{R}^d = \bigcup_i c_i$ into d-dimensional non-overlapping sets c_i. Such arrangements occur in many natural situations and depending on the situation the sets c_i might be called cells, crystals, regions, tiles, etc.

Many real-life tessellations are random. Random tessellations have been studied for a long time in stochastic geometry and a general theory has now been established, see e.g. Stoyan et al. (1987), Zähle (1988), Møller (1989), and Mecke et al. (1990). Usually the cells are assumed to be bounded and convex, and the aggregate of cells is locally bounded in the sense that the number of cells intersecting any bounded subset of $\mathbb{R}^d$ is finite. Then the cells become d-dimensional convex polytopes, i.e. bounded intervals (d = 1), convex polygons (d = 2), convex polyhedra (d = 3), etc. Of course, most interest in 'practice' concerns planar and spatial random tessellations.

Indeed there is a large variety of specific probabilistic models for random tessellations. Typically, the random mechanism is given by some stochastic process of simple geometrical objects which generate the tessellation in accordance to some rule. One such example is a process of lines in $\mathbb{R}^2$, which in an obvious way determines a random planar tessellation of $\mathbb{R}^2$, see Figure 1.1. Line-generated tessellations provide one of the mathematically most tractable class of models of tessellations, especially if the line process is Poisson, see e.g. Mecke et al. (1990) and the references therein. However, their importance for 'practical' applications seem to be somewhat limited.

In these lecture notes we shall consider another construction: Voronoi tessellations generated by point processes as described in Section 1.1. In the first part, Chapters 1-2, we discuss essentially non-random properties of Voronoi tessellations, i.e. the geometric structure of the cell aggregate when the realization of the associated point process is given; also

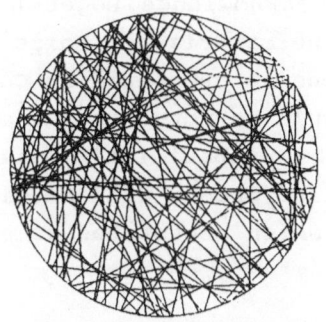

Figure 1.1. An example of a simulated Poisson line tessellation observed within a disc.

some background material on integral geometry is presented. The remaining Chapters 3-4, which constitute the major part, are devoted to the study of random Voronoi tessellations. Chapter 3 treats arbitrary stationary Voronoi tessellations using Palm measure theory. Chapter 4 concerns Poisson-Voronoi tessellations. The homogeneous Poisson process seems to be the only non-trivial stochastic model for which a reasonable collection of theoretical results for the associated Voronoi tessellation is derivable. For specificity and ease of exposition, we shall often restrict attention to planar and spatial Voronoi tessellations. However, most concepts and results hold as well in $\mathbb{R}^d$, and sometimes it will be more appropriate to state the definitions and results for arbitrary dimensions d.

The notes in no way attempt to cover all aspects of random Voronoi tessellations. Much deeper and fascinating results have had to be omitted, like e.g. David G. Kendall's shape theory of Poisson-Delaunay triangles, see Kendall (1989). Further, many results hold as well for more general models of tessellations. This is in particular the case for the mean value relations stated in Chapter 3 as established by Joseph Mecke and many others. Also I must acknowledge Roger E. Miles and Kenneth A. Brakke for most of the material on Poisson-Voronoi and Poisson-Delaunay tessellations as presented in Chapter 4.

Finally, it should be noted that the exposition will mainly concentrate on the probabilistic aspects of random Voronoi tessellations. Detailed proofs will be given with the exception of some easy proofs which are stated as exercises.

1.1. Definitions, assumptions, and characteristics

In this section we introduce the Voronoi and Delaunay tessellations and describe some of their geometric structure. More details are to be found in Section 2.1.

Consider a set $\Phi = \{x_i\}$ of points $x_i \in \mathbb{R}^d$ called nuclei. Each nucleus generates a cell

$$C(x_i|\Phi) = \{y \in \mathbb{R}^d \mid \|x_i - y\| \leq \|x_j - y\| \text{ for all } x_j \in \Phi\} \quad (1.1.1)$$

where $\|\cdot\|$ denotes Euclidean distance. Thus $C(x_i|\Phi)$ consists of all points which have x_i as nearest nucleus, see Figures 1.1.1-1.1.2. Equivalently,

$$C(x_i|\Phi) = \bigcap_{j:j\neq i} H(x_i, x_j) \quad (1.1.2)$$

where $H(x_i, x_j)$ is the closed halfspace

$$H(x_i, x_j) = \{y \in \mathbb{R}^d \mid (y - z_{ij}) \cdot (x_i - x_j) \geq 0\}$$

containing x_i and bounded by the bisecting hyperplane $G(x_i, x_j)$ of x_i and x_j, that is the hyperplane which contains the midpoint

$$z_{ij} = \tfrac{1}{2}(x_i + x_j)$$

and is perpendicular to the line through x_i and x_j. Here $\cdot$ is the usual inner product on $\mathbb{R}^d$. Thus $C(x_i|\Phi)$ is a closed convex set. Further, $C(x_i|\Phi) \cap C(x_j|\Phi) \subseteq G(x_i,x_j)$, so the cells are seen to have disjoint topological interiors. Consequently, the cells constitute a tessellation of $\mathbb{R}^d$ provided that any point $y \in \mathbb{R}^d$ has a nearest nucleus and the cells are of dimension d. This is easily seen to be the case if Φ is locally finite in the sense that the number of nuclei within any bounded subset of $\mathbb{R}^d$ is finite (see Exercise 1.1.1).

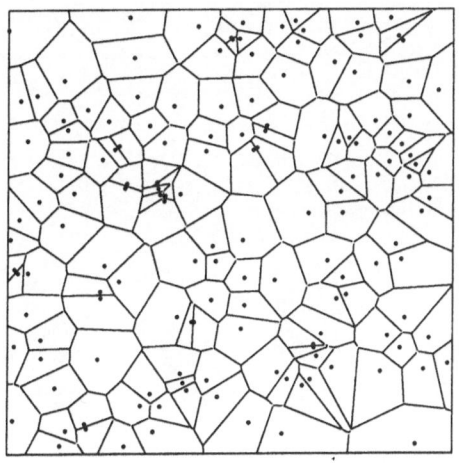

Figure 1.1.1. Planar Voronoi tessellation generated by a binomial process of 150 independent and uniformly distributed points on the unit square.

Definition 1.1.1. Suppose $\Phi = \{x_i\} \subset \mathbb{R}^d$ is locally finite. Then $C(x_i|\Phi)$ is called the Voronoi cell generated by the nucleus x_i. The aggregate of these cells constitutes the Voronoi tessellation $\mathcal{V}(\Phi)$ associated with Φ. □

Throughout these notes Φ is assumed to be a locally finite subset of $\mathbb{R}^d$ or equivalently a realization of a simple point process in $\mathbb{R}^d$, cf. Section 3.1. Note that this implies that Φ is either finite or at least countable. However, it does not ensure that the Voronoi cells become neither bounded nor convex polytopes; see Section 2.1 for further details.

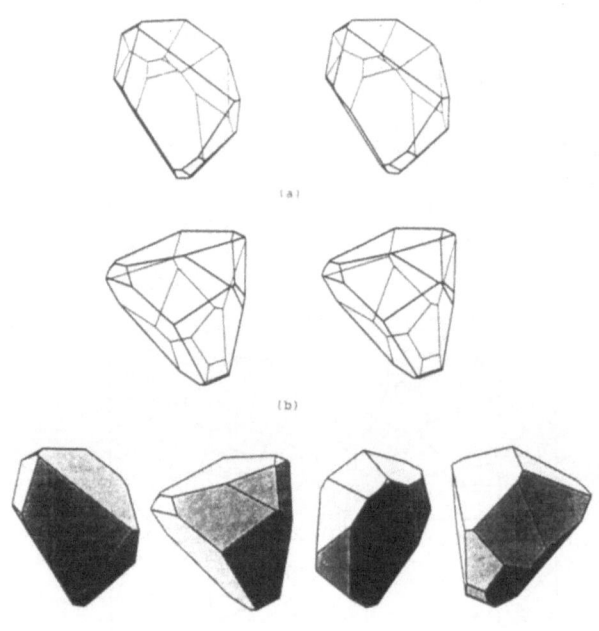

Figure 1.1.2. (a) Stereo pair of a random selected simulated Poisson-Voronoi cell. (b) The stereo pair after rotation with 90° about a fixed axis. (c) The cell shown after rotation with 0°, 90°, 180° and 270° about the fixed axis. (Taken from Møller et al. (1989)).

Henceforth we shall also assume that the nuclei are in general quadratic position, that is,

(a)   no   k+1   nuclei lie on a   k-1   dimensional affine subspace of $\mathbb{R}^d$,   k = 2,...,d,   and

(b)   no   d+2   nuclei lie on the boundary of a sphere.

In fact, any point process, which is absolutely continuous with respect to the usual Poisson point process defined on any (bounded or not) region $S \subseteq \mathbb{R}^d$,   generates realizations Φ which almost surely satisfy (a)-(b), cf. Section 4.1. General quadratic position of Φ exclude 'degenerate' cases like the regular lattice $\Phi = \mathbb{Z}^d$ which just generates congruent cubic Voronoi cells.

Notice that the intersection

$$F(x_0,\ldots,x_k|\Phi) = \bigcap_{i=0}^{k} C(x_i|\Phi) \qquad (1.1.3)$$

between   k+1   cells is included in the affine subspace

$$G(x_0,\ldots,x_k) = \{y \in \mathbb{R}^d | \|y-x_0\| = \ldots = \|y-x_k\|\} \qquad (1.1.4)$$

which is called a mathematical (d-k)-facet for   k = 1,...,d. It can be shown (see Exercise 2.1.1) that condition (a) is equivalent to assuming that any mathematical (d-k)-facet has dimension   d-k   for   k = 1,...,d.   Moreover, condition (b) is clearly equivalent to assuming that $G(x_0,\ldots,x_k) = \emptyset$ whenever $x_0,\ldots,x_k$ are distinct nuclei with   k > d+1.   In fact (a)-(b) and the local finiteness of Φ imply that (1.1.3) is either the empty set or of dimension   d-k   (see Proposition 2.1.1); in the non-void case $F(x_0,\ldots,x_k|\Phi)$ is called a (d-k)-facet. For convenience, cells are also called d-facets.

Since the facets are intersections between closed convex sets, the facets become closed and convex too. Thus, a 0-facet is the common vertex of exactly   d+1   cells, a 1-facet is the common edge of exactly   d   cells,...,and a (d-1)-facet is the interface between two neighbouring cells; see Figures 1.1.1 and 1.1.2. Note that a facet of dimension > 0 can be unbounded; for instance, an edge is either a bounded closed line segment ending in two vertices or a closed halfline ending in one vertex (provided that we have at least d+1 nuclei). If all cells are bounded, then the facets become bounded convex polytopes, i.e. points, line seg-ments, convex polygons, convex polyhedra, etc. (cf. Remark 2.1.2 and Proposition 2.1.4).

In the following, for ease of presentation, let us assume that Φ consists of at least d+1 nuclei. Then, $\mathcal{V}(\Phi)$ has at least one vertex (see Proposition 2.1.2). Define the closed d-dimensional simplex

$$T(x_0,\ldots,x_d) = \{ \sum_{i=0}^{d} \alpha_i x_i \,|\, 0 \le \alpha_i \le 1, \ \sum_{i=0}^{d} \alpha_i = 1 \} \qquad (1.1.5)$$

for any vertex $z_{0\ldots d} = F(x_0,\ldots,x_d|\Phi)$ of $\mathcal{V}(\Phi)$ (a k-dimen-sional simplex in $\mathbb{R}^d$ is the convex hull of k+1 points, so it is just a point (k=0), line segment (k=1), triangle (k=2), tetrahedron (k=3), etc.) These simplices defined for all vertices of $\mathcal{V}(\Phi)$ are called Delaunay cells, and it can be shown that they constitute a tessellation of $S = \text{conv}\{\Phi\}$, the closed convex hull of Φ, see Figure 1.1.3 and Proposition 2.1.5. Note that $z_{0\ldots d}$ is the center of the open ball $B(x_0,\ldots,x_d)$ which contains $x_0,\ldots,x_d$ in its topological boundary. It follows from (1.1.3) that $F(x_0,\ldots,x_d|\Phi) \ne \emptyset$ if and only if $B(x_0,\ldots,x_d) \cap \Phi = \emptyset$.

<u>Definition 1.1.2</u>. The aggregate of Delaunay cells $T(x_0,\ldots,x_d)$ defined for any d+1 nuclei $x_0,\ldots,x_d \in \Phi$ with $B(x_0,\ldots,x_d) \cap \Phi = \emptyset$ constitutes the Delaunay tessellation $\mathcal{T}(\Phi)$ associated with Φ. ◻

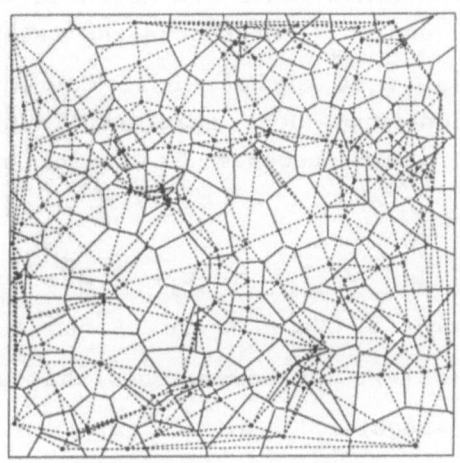

Figure 1.1.3. Same as Figure 1.1.1 but with the Delaunay tessellation shown (dashed lines).

Obviously, the edges of $\mathcal{T}(\Phi)$ join nuclei whose Voronoi cells are neighbours. The converse is also true: The line segment between the nuclei of any pair of neighbouring Voronoi cells is an edge of $\mathcal{T}(\Phi)$ (see Exercise 2.1.3). Consequently, using (1.1.2), there is a one-to-one correspondence between $\mathcal{V}(\Phi)$ and $\mathcal{T}(\Phi)$. In fact, they are dual in a graph theoretical sense as there is a one-to-one correspondence between the (d-k)-facets of $\mathcal{V}(\Phi)$ and the k-faces of $\mathcal{T}(\Phi)$, k = 0,..,d. Many algorithms for the construction of $\mathcal{V}(\Phi)$ is based on this duality.

The Delaunay tessellation contains various connectivity graphs, but here we shall consider only the Gabriel graph; for other examples see e.g. Aurenhammer (1991). Two nuclei x_i and x_j are said to be Gabriel neighbours if

$$z_{ij} = \tfrac{1}{2}(x_i + x_j) \in F(x_i, x_j | \Phi) \qquad (1.1.6)$$

that is, if their Voronoi cells are neighbours and the Delaunay edge joining x_i and x_j intersects the interface between the two Voronoi cells. Some authors call Gabriel neighbours for full Voronoi neighbours. Note that each nucleus has another nearest nucleus, which becomes a Gabriel neighbour to the nucleus.

__Exercise 1.1.1.__ If $\Phi = \{x_i\} \subset \mathbb{R}^d$ is locally finite prove that for any $y \in \mathbb{R}^d$ there exists a nucleus $x_i \in \Phi$ such that $y \in C(x_i|\Phi)$. Moreover, show that any cell $C(x_i|\Phi)$ is topologically regular, i.e., $C(x_i|\Phi) = \mathrm{cl}(\mathrm{int}\ C(x_i|\Phi))$ where 'cl' and 'int' denote topological closures and interior (notice that this result implies that $C(x_i|\Phi)$ has dimension d).

Hint: To prove that $C(x_i|\Phi) \subseteq \mathrm{cl}(\mathrm{int}\ C(x_i|\Phi))$, let $y \in C(x_i|\Phi)$ and consider $x = x_i + t(y - x_i)$ for $0 \leq t < 1$. Then, using the triangle inequality, show that $\|x - x_i\| < \|x - x_j\|$ for $j \neq i$, and so by the local finiteness of Φ that $x \in \mathrm{int}\ C(x_i|\Phi)$. Now, observe that this implies that $y \in \mathrm{cl}(\mathrm{int}\ C(x_i|\Phi))$. Finally, as $x_i \in \mathrm{int}\ C(x_i|\Phi)$ (why?), we find that $C(x_i|\Phi)$ has dimension d.

1.2. History and applications

The history of Voronoi tessellations goes back to Dirichlet (1850) who used a special form of the Voronoi tessellation in $\mathbb{R}^2$ and $\mathbb{R}^3$ in his study of positive definite quadratic forms. Later Voronoi (1908) generalized this in a more straightforward manner to higher dimensions. Since then Voronoi tessellations have been discovered in diverse scientific disciplines under many different names such as Dirichlet regions, Thiessen (1911) polygons, and Wigner and Seitz (1933) zones, although 'Voronoi tessellation' seems to be the most common name. The dual Delaunay tessellation was introduced by Voronoi (1908) for nuclei positioned at a regular lattice and was later extended by Delaunay (1934) to nuclei in general position. (Delaunay is a synonym used by the Soviet number theorist Boris Nikolaevitch Delone (1890-1980)).

Voronoi tessellations provide models for many natural phenomena or processes in crystallography, metallography, physics, astrophysics, biology, ecology, geology, geography, etc.; for examples and references, see e.g. Hinde and Miles (1980), Boots (1987), Stoyan et al. (1987), Aurenhammer (1991), and van de Weygaert (1991). This is mainly due to the nearest-nucleus definition of Voronoi cells which makes it possible to intepret planar Voronoi cells as territories, market areas, etc.; and spatial Voronoi cells as e.g. crystals generated by nuclei which start to grow at the same instant and at the same rate along all rays, but where the growth along a ray stops as it contacts another growing crystal. Statistical methods for modelling and analyzing Voronoi tessellations have been discussed in e.g. Stoyan and Hermann (1986), Hermann et al. (1989), Møller et al. (1989), Lorz (1990), and Stoyan and Stoyan (1990).

Mathematical and statistical applications of Voronoi and Delaunay tessellations include packing theory (Rogers, 1964), spatial interpolation of surfaces (Sibson, 1980a), analysis of spatial point patterns (Ripley, 1981; Baddeley and Møller, 1989), and in particular stochastic geometry where Voronoi tessellations form an important subject (Stoyan et al., 1987). Mainly Voronoi tessellations generated by a homogeneous Poisson process have been studied, the pioneering works are due to Meijering (1953) and Gilbert (1962), while later contributions include Miles (1970, 1974, 1984), Hinde and Miles (1980), Miles and Maillardet (1982), Quine and Watson (1984), Møller (1989), and many others.

Currently, Voronoi tessellations are attracting much research interest in computer science, especially within computational geometry. For a recent survey, see Aurenhammer (1991), which contains numerous references and describes various algorithms for the construction of Voronoi and Delaunay tessellations.

1.3. Related tessellations

The definition of Voronoi cells can be generalized in many ways giving rise to sometimes more realistic but also more analytically complicated models of tessellations. In this section we give a review on some of these constructions.

A more natural model for crystal growth is obtained by considering nuclei $\{x_i\}$ which start to grow at different times $\{t_i\}$. Assuming that the nuclei grow at a constant rate $v > 0$ along all rays until they come into contact, crystals

$$C(x_i|\Phi) = \{y \in \mathbb{R}^d \mid T_i(y) \leq T_j(y) \text{ for all } j \neq i\}, \quad x_i \in \Phi \ , \quad (1.3.1)$$

are generated, where

$$T_i(y) = t_i + \|x_i - y\| / v$$

is the time x_i reaches y, see Figure 1.3.1. The non-empty crystals constitute a tessellation of hyperbolically star-shaped cells which are not necessarily convex. This model, due to Johnson and Mehl (1939), includes the Voronoi tessellation as the special case where all nuclei start growing simultaneously. A survey on random Johnson-Mehl tessellations is presented in Møller (1992).

A further generalization is obtained by allowing different growth rates $v_i > 0$. Then the crystals are not necessarily simply connected because crystals can be entirely contained

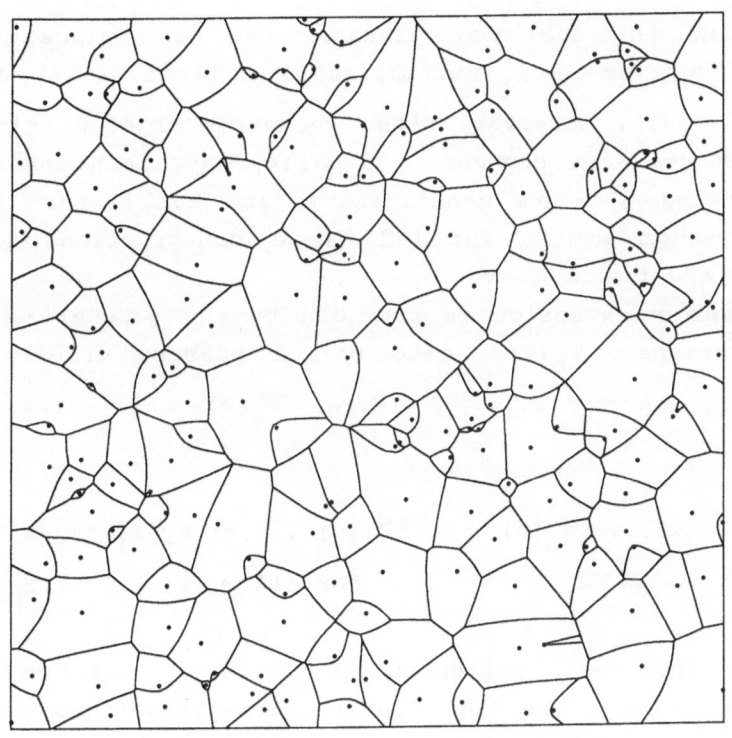

Figure 1.3.1. Johnson-Mehl tessellation generated by 300 nuclei within the unit square. To each nucleus is attached an exponentially distributed waiting time for the start of growth (mean waiting time = 100). These waiting times are mutually independent and independent of the positions of the nuclei, which are generated by a binomial process. Only those nuclei which generate non-empty crystals are shown (total = 222).

within other crystals. In the simplest case where all birth-times t_i agree, but all growth rates v_i disagree, the cell boundaries become spherical. This corresponds to assign a weight v_i to each of nucleus x_i and replace the Euclidean distance $\|x_i - y\|$ in the definition (1.3.1) by the weighted distance $\|x_i - y\| / v_i$. Other weighting schemes have been proposed and the Euclidean distance can be replaced by other metrics such as the L_p-metric, see e.g. Boots (1987) and Aurenhammer (1991). Moreover, other geometric objects than points, for instance line segments and balls, have been considered as the growing elements generating cells. But present knowledge about random models for all these constructions is rather limited and fragmentary.

Another extension is provided by the generalized Voronoi tessellations $\mathscr{V}_n(\Phi)$ also called nearest order-n Voronoi diagrams, $n = 1, 2, \ldots$. Here $\mathscr{V}_n(\Phi) = \{C(x_1, \ldots, x_n | \Phi) \neq \emptyset\}$ with

$$C(x_1, \ldots, x_n | \Phi) = \{y \in \mathbb{R}^d \mid i = 1, \ldots, n: \|y - x_i\| \leq \|y - x_j\|$$
$$\text{for all } x_j \in \Phi \setminus \{x_1, \ldots, x_n\}\}$$

defined for all configurations of n distinct nuclei $x_1, \ldots, x_n \in \Phi$. The tessellation is well-defined as Φ is locally finite. If Φ is finite, $\Phi = \{x_0, \ldots, x_n\}$ say, $\mathscr{V}_n(\Phi)$ is called the farthest nuclei tessellation because $C(x_1, \ldots, x_n | \Phi)$ consists of all points farthest from the nucleus x_0. Especially, $\mathscr{V}_1(\Phi) = \mathscr{V}(\Phi)$. Sibson (1980b) has shown that if $x_0 \in \Phi$ generates a bounded Voronoi cell with neighbouring nuclei $x_1, \ldots, x_k$, then x_0 can be written as the convex combination $x_0 = \sum_{i=1}^{k} \alpha_i x_i$ with

$$\alpha_i = |C(x_0 | \Phi) \cap C(x_0, x_i | \Phi)|_d / |C(x_0 | \Phi)|_d$$

being the relative volume of that part of $C(x_0 | \Phi)$ which is closer to x_i than any other nucleus of $\Phi \setminus \{x_0, x_i\}$. This formula has application in spatial interpolation and smoothing. A few properties of random generalized Voronoi tessellations are known mainly due to Miles (1970) and Miles and Maillardet (1982). The latter paper contains illuminating illustrations of $\mathscr{V}_n(\Phi)$ for $n = 4, 16, 64$, and 256 when Φ is a Poisson process, in which case it is observed that the cells tend to be more circular as n increases.

Still another possibility is to define Voronoi tessellations on other spaces than $\mathbb{R}^d$. Voronoi tessellations defined on the torus are used in order to eliminate edge effects; a torus is a square with its opposite edges identified. Also Voronoi tessellations defined on the sphere and on more abstract spaces have been studied.

Exercise 1.3.1. Show that $\mathscr{V}_n(\Phi)$ is a tessellation with convex topological regular cells.

Hint: Observe that

$$C(x_1, \ldots, x_n | \Phi) = \bigcap_{i=1}^{n} C(x_i | \Phi \setminus \{x_1, \ldots, x_{i-1}, x_{i+1}, \ldots, x_n\}) .$$

2. Geometrical properties and other background material

The concepts and results presented in this chapter are frequently used in later chapters.

Suppose we are given a locally finite set $\Phi = \{x_i\} \subset \mathbb{R}^d$ of nuclei which are in general quadratic position, cf. conditions (a)-(b) in Section 1.1. Then the system of facets of the Voronoi tessellation $\mathcal{V}(\Phi)$ possesses many desirable geometrical and topological properties as discussed in Section 2.1 below. In Section 2.2 we describe which characteristics of the facets become relevant if certain natural invariance conditions should be fulfilled.

2.1. On the geometric structure of Voronoi and Delaunay tessellations

In this section we establish some of the geometrical and topological properties of Voronoi and Delaunay tessellations which are relevant for our purposes.

Lemma 2.1.1. For $1 \leq k \leq d$ and any $k+1$ nuclei $x_0, \ldots, x_k \in \Phi$, let $L_{d-k} = L(x_0, \ldots, x_k)^\perp$ denote the orthogonal complement to the k-dimensional linear subspace $L_k = L(x_0, \ldots, x_k)$ spanned by $x_1-x_0, \ldots, x_k-x_0$. Further, let $z_{0\ldots k} = z(x_0, \ldots, x_k)$ denote the center of the sphere in x_0+L_k which contains $x_0, \ldots, x_k$. Then the mathematical (d-k)-facet (1.1.4) is given by

$$G(x_0, \ldots, x_k) = z_{0\ldots k} + L_{d-k} \qquad (2.1.1)$$

that is the (d-k)-dimensional affine subspace through $z_{0\ldots k}$ and orthogonal to L_k. □

Proof. Note that condition (a) ensures that L_k is k-dimensional and $z_{0...k}$ exists. By the definition (1.1.4),

$$y \in G(x_0, \ldots, x_k) \iff \|y-x_0\| = \ldots = \|y-x_k\|$$

$$\iff \|y\|^2 - 2x_0 \cdot y + \|x_0\|^2 = \ldots = \|y\|^2 - 2x_k \cdot y + \|x_k\|^2$$

so clearly $z_{0...k} \in G(x_0, \ldots, x_k)$ and

$$y \in G(x_0, \ldots, x_k) \iff \begin{bmatrix} (x_1 - x_0)^* \\ \vdots \\ (x_k - x_0)^* \end{bmatrix} y = \frac{1}{2} \begin{bmatrix} \|x_1\|^2 - \|x_0\|^2 \\ \vdots \\ \|x_k\|^2 - \|x_0\|^2 \end{bmatrix} \qquad (2.1.2)$$

where $*$ denotes transposition. This proves (2.1.1). □

Proposition 2.1.1. Let the situation be as in Lemma 2.1.1. Then

$$F(x_0, \ldots, x_k | \Phi) = \text{cl}(\text{rel int } F(x_0, \ldots, x_k | \Phi)) \qquad (2.1.3)$$

where 'rel int' means relative interior with respect to the topology on $G(x_0, \ldots, x_k)$ identified with $\mathbb{R}^{d-k}$. Thus any facet is topological regular, and it follows from (1.1.3) that each k-facet lies in the relative boundaries of $\begin{bmatrix} d-k+1 \\ 1-k \end{bmatrix}$ l-facets and contains in its relative boundary a certain number of j-facets, $0 \leq j \leq k \leq l \leq d$. □

Proof. As $F(x_0, \ldots, x_k | \Phi)$ is closed, it suffices to prove that $F(x_0, \ldots, x_k | \Phi) \subseteq \text{cl}(\text{rel int } F(x_0, \ldots, x_k | \Phi))$. This is trivial if $k = d$, because then $F(x_0, \ldots, x_k | \Phi)$ is either empty or just a vertex.

Suppose $1 \leq k < d$ and $y \in F(x_0, \ldots, x_k | \Phi)$, and let $x_0, x_1, \ldots, x_m \in \Phi$ denote those nuclei with $\|x_0 - y\| = \|x_1 - y\|$ $= \ldots = \|x_m - y\|$. From condition (b) we have that $k \leq m \leq d$. Using (1.1.3)-(1.1.4) and the local finiteness of Φ we find a small open ball $B(y, \epsilon) \subset \mathbb{R}^d$ with center y and radius ϵ such that

$$B(y, \epsilon) \cap F(x_0, \ldots, x_k | \Phi)$$

$$= \{x \mid \forall j \notin \{0, \ldots, k\} : \|x_0 - x\| \leq \|x_j - x\|\} \cap B(y, \epsilon) \cap G(x_0, \ldots, x_k)$$

$$= \bigcap_{j=k+1}^{m} H(x_0, x_j) \cap B(y, \epsilon) \cap G(x_0, \ldots, x_k)$$

$$\supset \bigcap_{j=k+1}^{m} (H(x_0, x_j) \backslash G(x_0, x_j)) \cap B(y, \epsilon) \cap G(x_0, \ldots, x_k) \quad (2.1.4)$$

(taking $\bigcap_{j=k+1}^{m} (H(x_0, x_j) \backslash G(x_0, x_j)) = \mathbb{R}^d$ if $k = m$). Recall that $H(x_0, x_j) \backslash G(x_0, x_j)$ is an open halfspace bounded by the hyperplane $G(x_0, x_j)$ which contains y. Now, by Lemma 2.1.1 and letting $y = z_{0 \ldots k} + y^{d-k}$ with $y^{d-k} \in L^{d-k}$, we have that

$$B(y, \epsilon) \cap G(x_0, \ldots, x_k) = z_{0 \ldots k} + [B(y^{d-k}, \epsilon) \cap L_{d-k}]$$

which is seen to be a $(d-k)$-dimensional ball within $G(x_0, \ldots, x_k)$. Further, this ball has center y and radius ϵ, and so its intersection with the $m-k$ ($\leq d-k$) open halfspaces in (2.1.4) becomes a relative open set within $G(x_0, \ldots, x_k)$ which contains y in its boundary; see also Figure 2.1.1. Consequently, $y \in cl(\text{rel int } F(x_0, \ldots, x_k | \Phi))$ and the proof is completed. $\quad \square$

18

Remark 2.1.1. Recall that Voronoi cells (d-facets) are also topological regular, cf. Exercise 1.1.1. □

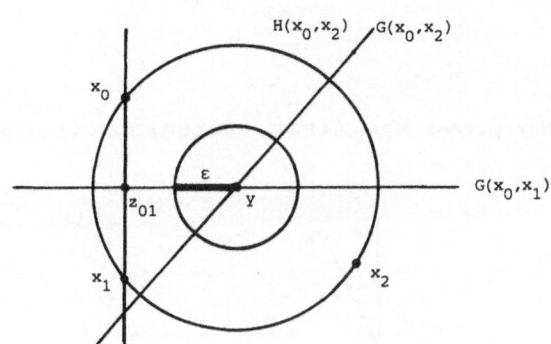

Figure 2.1.1. Illustration to the proof of Proposition 2.1.1 in the case d = 2, k = 1, m = 2. The thick line segment in the disc B(y,ε) contains the intersection (2.1.4).

Proposition 2.1.2. The set of facets is locally finite in the sense that the number of k-facets intersecting any bounded subset of $\mathbb{R}^d$ is finite, k = 0,1,...,d. Moreover, $\mathcal{V}(\Phi)$ contains facets of any dimension k = 0,1,...,d, provided card(Φ) ≥ d+1. □

Proof. See Exercise 2.1.2. □

Proposition 2.1.3. A Voronoi cell $C(x_0|\Phi)$ is bounded in $\mathbb{R}^d$ if and only if

$$\forall u \in S^{d-1} \exists x_1 \in \Phi\backslash\{x_0\}: (x_1-x_0) \cdot u > 0 \qquad (2.1.5)$$

where $S^{d-1} = \{u \in \mathbb{R}^d | \|u\|=1\}$ denotes the unit sphere. If

$C(x_0|\Phi)$ is bounded, then it has a finite number of neighbouring cells $C(x_1|\Phi),\ldots,C(x_n|\Phi)$, say, in which case

$$C(x_0|\Phi) = \bigcap_{i=1}^{n} H(x_0,x_i) \ . \qquad (2.1.6)$$

□

Proof. To prove the first statement consider the following figure:

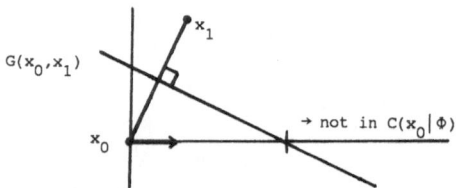

If $C(x_0|\Phi)$ is bounded, Proposition 2.1.2 implies that $C(x_0|\Phi)$ is intersected by a finite number of Voronoi cells. Finally, (2.1.6) follows from (1.1.2) and using the local finiteness of Φ. □

Remark 2.1.2. Recall that a convex polytope in $\mathbb{R}^d$ is a finite intersection between closed halfspaces of $\mathbb{R}^d$. We shall call it a k-polytope if it is bounded and of dimension k, so a k-polytope becomes just a point, line segment, convex polygon or convex polyhedron for $k = 0,1,2,3$, respectively. The boundary of a d-polytope

$$X_d = \bigcap_{i=1}^{n} H_i$$

defined by the closed halfspaces H_i, $i = 1,\ldots,n$, consists of a hierarchy of (d-k)-facets, $k = 0,\ldots,d$; a (d-k)-face is simply a (d-k)-polytope of the form

$$X_{d-k} = \bigcap_{i=1}^{k} G_i \cap \bigcap_{j=k+1}^{n} H_j$$

where G_i is the hyperplane which bounds H_i. For further details on polytopes, see e.g. Grünbaum (1967). □

Proposition 2.1.4. Suppose the Voronoi cell $C(x_0|\Phi)$ is bounded. Then $C(x_0|\Phi)$ becomes a d-polytope and its k-faces coincide with those k-facets of $\mathcal{V}(\Phi)$ which are included in $C(x_0|\Phi)$, $k = 0,\ldots,d$. Moreover, if all cells of $\mathcal{V}(\Phi)$ are bounded, the facets of $\mathcal{V}(\Phi)$ and the faces of the Voronoi cells $C(x_i|\Phi)$, $x_i \in \Phi$, coincide. □

Proof. It follows from (2.1.6) that $C(x_0|\Phi)$ is a d-polytope. By (1.1.3) and (1.1.4), we find that

$$F(x_0,\ldots,x_k|\Phi)$$

$$= G(x_0,\ldots,x_k) \cap C(x_0|\Phi)$$

$$= \bigcap_{i=1}^{k} G(x_0,x_i) \cap \bigcap_{j=1}^{n} H(x_0,x_j)$$

$$= \bigcap_{i=1}^{k} G(x_0,x_i) \cap \bigcap_{j=k+1}^{n} H(x_0,x_j) \qquad (2.1.7)$$

where we have used (2.1.6) and the fact that $G(x_0,x_i) \subset H(x_0,x_i)$ for $i = 1,\ldots,k$. This proves the remaining part of the proposition. □

We turn now to the Delaunay tessellation $\mathcal{T}(\Phi)$ assuming that Φ contains at least $d+1$ nuclei. Let $S = \mathrm{conv}\{\Phi\}$

denote the convex hull of Φ, that is, the smallest convex set containing Φ, so

$$S = \{ \sum_{i=1}^{k} \alpha_i x_i \mid \sum_{i=1}^{k} \alpha_i = 1,\ 0 \leq \alpha_i \leq 1,\ x_i \in \Phi,\ k=1,2,\ldots \} \qquad (2.1.8)$$

Proposition 2.1.5. $\mathcal{T}(\Phi)$ constitutes a tessellation of S.

□

Proof. According to Definition 1.1.2 we shall show (i) that

$$S = \cup\{T(x_0,\ldots,x_d) \mid B(x_0,\ldots,x_d) \cap \Phi = \emptyset,\ x_0,\ldots,x_d \in \Phi\} \quad (2.1.9)$$

and (ii) that for any two different sets $\{x_0,\ldots,x_d\} \subseteq \Phi$ and $\{x_0',\ldots,x_d'\} \in \Phi$ with

$$B(x_0,\ldots,x_d) \cap \Phi = B(x_0',\ldots,x_d') \cap \Phi = \emptyset \qquad (2.1.10)$$

the Delaunay cells $T(x_0,\ldots,x_d)$ and $T(x_0',\ldots,x_d')$ have disjoint topological interiors. The statement (ii) is easily shown as $T(x_0,\ldots,x_d) \subset B(x_0,\ldots,x_d)$ and (2.1.10) implies that if $B(x_0,\ldots,x_d) \cap B(x_0',\ldots,x_d') \neq \emptyset$ then $x_0,\ldots x_d$ and $x_0',\ldots x_d'$ must lie on opposite sides of the hyperplane containing the sphere $\partial B(x_0,\ldots,x_d) \cap \partial B(x_0',\ldots,x_d')$ where ∂ denotes topological boundary. For (i), using (1.1.5) and (2.1.8), it suffices to show that the left hand side is included in the right hand side of (2.1.9). For this we proceed as in Rogers (1964, pp. 92-94).

Suppose that $a \in S$ is not contained in any $\mathcal{T}(\Phi)$-cell. Then, as Φ is locally finite, there is a small ball $B(a,\epsilon) \subset S$ with no point in any $\mathcal{T}(\Phi)$-cell. Let b be any point contained in the interior of some $\mathcal{T}(\Phi)$-cell (at least one cell exists, cf. Proposition 2.1.2). The union of all lines from b to any $(d-2)$-face of a $\mathcal{T}(\Phi)$-cell is a simplex of dimension $\leq$ $d-1$. Further, there is only countable many $(d-2)$-faces. Thus there is a point $c \in B(a,\epsilon)$ such that the closed line segment $[b,c]$ intersects no $(d-2)$-faces of the $\mathcal{T}(\Phi)$-cells.

Let $p \in [b,c]$ be the closest point to b so that the halfopen line segment $(p,c]$ has no point in any $\mathcal{T}(\Phi)$-cell. Then $p \neq c$ and p is an inner point of a $(d-1)$-face $T_{1...d}$ $= \mathrm{conv}\{x_1,\ldots,x_d\}$ of some $\mathcal{T}(\Phi)$-cell $T_{0...d} = \mathrm{conv}\{x_0,\ldots,x_d\}$.

Let $H = z_{1...d} + L_{d-1}$ be the hyperplane containing $x_1,\ldots,x_d$ and let $1 = z_{1...d} + L_1$ be the line of points equidistant from $x_1,\ldots x_d$, cf. Lemma 2.1.1; see also Figure 2.1.2. Then the edge $e = F(x_1,\ldots,x_d|\Phi)$ ends at the vertex $z_{0...d} = F(x_0,\ldots,x_d|\Phi)$ and is contained in 1. Further, x_0 and $z_{0...d}$ are both on the opposite side to that side of H which contains c. Furthermore, the points of $[z_{0...d},$ $z_{1...d}]$, which are sufficiently close to $z_{0...d}$ and on the same side of H as c, lie on the edge e. But $[b,c] \subset S$, as $b,c \in S$ where S is convex, so there exist nuclei from $\Phi \backslash \{x_0,\ldots,x_d\}$ which are on the same side of H as c. This implies that e has another endpoint, which becomes the $\mathcal{V}(\Phi)$-vertex of some $\mathcal{T}(\Phi)$-cell $\acute{T} = \mathrm{conv}\{x_1,\ldots,x_{d+1}\}$.

Since T and $\acute{T}$ have disjoint interiors, x_0 and x_{d+1} lie on opposite sides of H and the inner points of $T_{1...d}$ are contained in the interior of $T \cup \acute{T}$. Especially, p is an inner point of $T \cup \acute{T}$, and so $(p,c]$ has a point contained in $\acute{T}$. This contradiction proves that each point of S is contained in some $\mathcal{T}(\Phi)$-cell. □

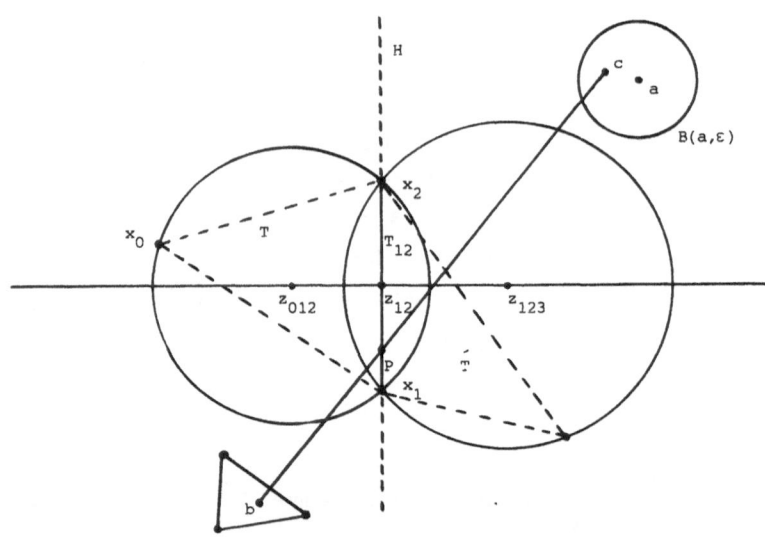

Figure 2.1.2. Illustration to the proof of Proposition 2.1.4 when $d = 2$. The edge e has endpoints z_{012} and z_{123}.

 Remark 2.1.3. According to Proposition 2.1.3, all Voronoi cells of $\mathscr{V}(\Phi)$ are bounded if any only if the condition

(c) $\quad \forall u \in S^{d-1} \ \forall x_0 \in \Phi \ \exists x_1 \in \Phi \backslash \{x_0\}: \ (x_1 - x_0) \cdot u > 0$

holds, in which case we have that $\mathscr{T}(\Phi)$ is a tessellation of $S = \text{conv}\{\Phi\} = \mathbb{R}^d$. □

 The k-faces of a Delaunay cell $T(x_0, \ldots, x_d)$ are simply given by the k-dimensional simplicies

$$T(y_0, \ldots, y_k) = \text{conv}\{y_0, \ldots, y_k\}$$

for all configurations $\{y_0, \ldots, y_k\} \subseteq \{x_0, \ldots, x_d\}$ of $k+1$

distinct nuclei, $k = 0,\ldots,d$. As in the definition of Voronoi facets, let us call a non-void intersection between any Delaunay cells a Delaunay k-facet if its dimension is k.

Proposition 2.1.6. The k-facets of $\mathcal{T}(\Phi)$ coincide with the k-faces of the Delaunay cells of $\mathcal{T}(\Phi)$, $k = 0,\ldots,d$. □

Proof. Consider two $\mathcal{T}(\Phi)$ cells $T(x_0,\ldots,x_d)$ and $T(x_0',\ldots,x_d')$. Then $\{x_0,\ldots,x_d\}$ and $\{x_0',\ldots,x_d'\}$ are on opposite sides of a hyperplane, so

$$T(x_0,\ldots,x_d) \cap T(x_0',\ldots,x_d') = T(\{x_0,\ldots,x_d\} \cap \{x_0',\ldots,x_d'\}) \ .$$

Hence, by induction, we obtain for any $\mathcal{T}(\Phi)$ cells $T(x_{i0},\ldots,x_{id})$, $i = 1,\ldots,n$, that

$$\bigcap_{i=1}^{n} T(x_{i0},\ldots,x_{id}) = T(\bigcap_{i=1}^{n} \{x_{i0},\ldots,x_{id}\})$$

whereby the proposition follows. □

In the remaining part of this section we consider only the planar case $d = 2$.

Remark 2.1.4. Let $d = 2$. Then Proposition 2.1.5 states that $\mathcal{T}(\Phi)$ is a triangulation of $S = \mathrm{conv}\{\Phi\}$. Thus, if $n = \mathrm{card}(\Phi) < \infty$, Euler's formula gives

$$t-s+n = 1 \qquad\qquad (2.1.11)$$

where t,s,n are the number of Delaunay triangles, edges, and vertices, respectively. Furthermore, using induction on n one can prove that

$$t = 2(n-1) - n_h \tag{2.1.12}$$

where n_h denotes the number of vertices on the boundary of S. Thus combining (2.1.11) and (2.1.12),

$$s = 3(n-1) - n_h . \tag{2.1.13}$$

□

Remark 2.1.5. Let $d = 2$. Each pair of adjacent Delaunay triangles forms a quadrilateral, and the smallest angle of such two Delaunay triangles is always greater than the smallest angle of the other two triangles of the quadrilateral, see Figure 2.1.3. Sibson (1978) has shown that the Delaunay tessellation uniquely possesses this property of all triangulations with vertices given by Φ. □

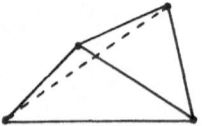

Figure 2.1.3. The thick diagonal of the quadrilateral is the common Delaunay edge.

Proposition 2.1.7. Let $d = 2$ and consider a vertex $z_{012} = F(x_0, x_1, x_2 | \Phi)$ of $\mathcal{V}(\Phi)$. Define the angles $\alpha, \alpha', \beta, \beta'$ as in the following figure, where $G_{ij} = G(x_i, x_j)$ denotes the bisecting line of the Voronoi cells $C(x_i | \Phi)$ and $C(x_j | \Phi)$:

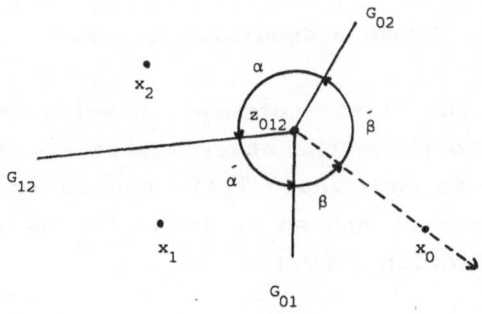

Then the 'direction line' from the vertex z_{012} to the nucleus x_0 (the dashed halfline) with respect to the edges of $C(x_0|\Phi)$ meeting at z_{012} are given by

$$\beta = \pi - \alpha \quad \text{and} \quad \beta' = \pi - \alpha' \; . \qquad (2.1.14)$$

□

Proof. See Exercise 2.1.4.   □

Remark 2.1.6.   The boundary of a Voronoi cell does not determine its nucleus. However the direction lines of the vertices meet at the nucleus. This can be used to recovering the nuclei given the Voronoi edges. Van de Weygaert (1991) used this fact to establish that a planar section through a spatial Voronoi tessellation does not in general generate a planar Voronoi tessellation. Moreover, if a planar tessellation has to be approximated by a Voronoi tessellation, then the nuclei might be obtained by minimizing some measure of distance between the nuclei and the generating lines determined by the original tessellation.   □

Exercise 2.1.1.   Show that condition (a) is equivalent to assume that any mathematical (d-k)-facet has dimension (d-k), $k = 1, \ldots, d$.

Hint: See the proof of Lemma 2.1.1.

Exercise 2.1.2. Prove Proposition 2.1.2.

Hints: (i) To prove the first statement it suffices to consider
the d-facets. (ii) To prove the other statement, use condition
(a) and Lemma 2.1.1 to show that $\mathcal{T}(\Phi)$ has no (d-k)-facets if
it has no (d-k-1)-facets, and so by induction no (d-1)-facets;
this gives a contradiction (why?).

Exercise 2.1.3. For any two nuclei $x_0, x_1 \in \Phi$ show that
$F(x_0, x_1 | \Phi) \neq \emptyset$ implies that the line segment with endpoints
x_0 and x_1 becomes an edge of some $\mathcal{T}(\Phi)$-simplex.

Hint: Use the hint (ii) to Exercise 2.1.2.

Exercise 2.1.4. Prove Proposition 2.1.7 by considering
the following figure:

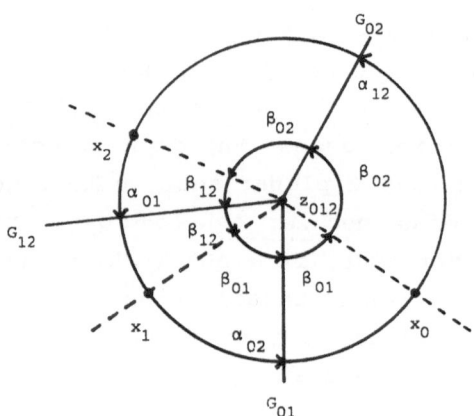

2.2. Short diversion into integral geometry

This section contains a brief account on some concepts and
results from integral geometry which are relevant for our pur-

poses. For references and a recent introduction to integral geometry, see Mecke et al. (1990). For specificity and simplicity we restrict attention to the planar and spatial cases.

Let $\mathcal{L}_k^{(d)}$ denote the space of k-dimensional linear subspaces of $\mathbb{R}^d$. On $\mathcal{L}_k^{(d)}$ a unique probability measure dL_k exists which is invariant under rotations in $\mathbb{R}^d$. Of course, dL_0 and dL_d are just the point measures concentrated on $\mathcal{L}_0^{(d)} = \{0\}$ and $\mathcal{L}_d^{(d)} = \{\mathbb{R}^d\}$, respectively. Furtermore, let v_d denote the surface measure on $\mathbb{R}^d$, let $S_+^{d-1} \subset S^{d-1}$ be the upper halfcircle (d=2) or hemisphere (d=3), and set $c_d = v_d(S_+^{d-1})$, i.e. $c_2 = \pi$ and $c_3 = 2\pi$. We can identify $L_1 \in \mathcal{L}_1^{(d)}$ by its direction $u \in S_+^{d-1}$. Then

$$dL_1 = v_d(du)/c_d , \qquad d = 2,3 . \qquad (2.2.1)$$

Similarly, when $d = 3$ and $L_2 \in \mathcal{L}_2^{(3)}$, we can identify L_2 by the unit vector $\{u\} = S_+^2 \cap L_2^{\perp}$, and so

$$dL_2 = v_3(du)/c_3 , \qquad d = 3 . \qquad (2.2.2)$$

Specifically, if $d = 2$, then $u = (\cos\theta, \sin\theta)$ with $\theta \in [0,\pi)$ and

$$v_2(du) = d\theta , \qquad d = 2 , \qquad (2.2.3)$$

while for $d = 3$, $u = (\cos\theta \sin\phi, \sin\theta \sin\phi, \cos\phi)$ with $0 \leq \theta < 2\pi$, $0 \leq \phi \leq \pi$, and

$$v_3(du) = \sin\phi \, d\theta d\phi , \qquad d = 3 , \qquad (2.2.4)$$

see also Figure 2.2.1.

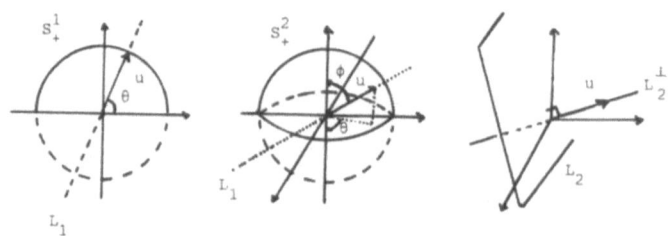

Figure 2.2.1. Left: d = 2. Middle and right: d = 3.

Let $\mathcal{G}_k^{(d)}$ denote the space of k-dimensional affine sub-spaces of $\mathbb{R}^d$. For short, we call $G_k \in \mathcal{G}_k^{(d)}$ a k-flat, so G_0 is just a point, G_1 a line, G_2 a plane, and $G_d = \mathbb{R}^d$. The Lebesgue measure on a k-flat is denoted dy^k, where the upper index indicates the dimension. Each k-flat G_k is in one-to-one correspondence with the parallel k-space $L_k \in \mathcal{L}_k^{(d)}$ and a unique point $y^{d-k} \in L_k^{\perp}$, the orthogonal complement to L_k, so that

$$G_k = y^{d-k} + L_k \qquad (2.2.5)$$

The measure on $\mathcal{G}_k^{(d)}$ given by

$$dG_k = dy^{d-k}\, dL_k \qquad (2.2.6)$$

is invariant under motions in $\mathbb{R}^d$, i.e. translations and rotations in $\mathbb{R}^d$ (see Exercise 2.2.1). In fact, dG_k is up to multiplication with a positive constant, the unique measure on $\mathcal{G}_k^{(d)}$ which is invariant under motions.

Now, let $\mathcal{K}_d$ be the class of convex compact subsets of $\mathbb{R}^d$. For d = 1,2,3, k = 0,...,d, and $K \in \mathcal{K}_d$ define

$$v_k^{(d)}(K) = c_k^{(d)} \int \chi(K \cap G_{d-k}) dG_{d-k} \qquad (2.2.7)$$

where

$$\chi(K) = \begin{cases} 1 & \text{if } K \neq \emptyset \\ 0 & \text{else} \end{cases}$$

and

$$c_k^{(d)} = \begin{cases} \pi , & d = 2, \quad k = 1 \\ 4 , & d = 3, \quad k = 2 \\ 1 & \text{else} \end{cases}$$

Note that $v_k^{(d)}$ is a non-negative functional on $\mathcal{K}_d$ which is invariant under motions in $\mathbb{R}^d$, monotone in the sense that $v_k^{(d)}(K_1) \leq v_k^{(d)}(K_2)$ whenever $K_1 \subseteq K_2$, and additive in the sense that

$$v_k^{(d)}(K_1 \cup K_2) = v_k^{(d)}(K_1) + v_k^{(d)}(K_2) - v_k^{(d)}(K_1 \cap K_2)$$

provided that $K_1 \cup K_2 \in \mathcal{K}_d$ ($v_k^{(d)}$ is proportional to the so-called $(d-k)$'th Quermass integral). According to a theorem due to Hadwiger (1957, pp. 221-225) any non-negative functional on $\mathcal{K}_d$, which is invariant under motions in $\mathbb{R}^d$, monotone, and additive becomes a linear combination of $v_0^{(d)}, \dots, v_d^{(d)}$ with non-negative coefficients.

Inserting (2.2.6) into (2.2.7) and using (2.2.2)-(2.2.3) we find that

$$v_k^{(d)}(K) = c_k^{(d)} \int |\pi_{L_k}(K)|_k dL_k \qquad (2.2.8)$$

where π_{L_k} denotes the orthogonal projection of $\mathbb{R}^d$ on L_k

and $|\cdot|_k$ means k-dimensional content, i.e. Lebesgue measure on L_k when $k > 0$ and counting measure when $k = 0$. In fact

$$V_0^{(d)}(K) = \chi(K) , \qquad d = 1,2,3 \qquad\qquad (2.2.9)$$

$$V_1^{(1)}(K) = L(K) = \text{length of } K \qquad\qquad (2.2.10)$$

$$V_1^{(2)}(K) = U(K) = \text{perimeter of } K \qquad\qquad (2.2.11)$$

$$V_2^{(2)}(K) = A(K) = \text{area of } K \qquad\qquad (2.2.12)$$

$$V_1^{(3)}(K) = \bar{b}(K) = \text{mean breadth of } K \qquad\qquad (2.2.13)$$

$$V_2^{(3)}(K) = S(K) = \text{surface area of } \partial K \qquad\qquad (2.2.14)$$

$$V_3^{(3)}(K) = V(K) = \text{volume of } K \qquad\qquad (2.2.15)$$

where the term 'mean breadth' is used because

$$\bar{b}(K) = \int |\pi_{L_1}(K)|_1 \, dL_1 .$$

Formulae (2.2.9)-(2.2.15) except (2.2.11) and (2.2.14) are all trivial consequences of (2.2.8). Formulae (2.2.11) and (2.2.14) can be verified directly by first considering the case in which K is a polytope and next using that any convex compact set can be expressed as the limit of a sequence of convex polytopes. Alternatively, Hadwiger's theorem can be used (see Exercise 2.2.2). We shall refer to $V_k^{(d)}$ as the k-dimensional content on $\mathcal{K}_d$.

We can identify a given k-flat $G_k \in \mathcal{G}_k^{(d)}$ by $\mathbb{R}^k$. Then

$v_1^{(k)}$ denotes the 1-dimensional content on $G_k \cap \mathfrak{K}_d$, the convex compact subsets of G_k. This is related to $v_1^{(d)}$ by

$$L(K) = \begin{cases} \frac{1}{2} U(K), & d = 2 \\ 2 \bar{b}(K), & d = 3 \end{cases}$$

if K is a line segment (the case $k = 1 = 1$), and

$$U(K) = 4 \bar{b}(K), \qquad A(K) = \frac{1}{2} S(K), \qquad \text{if} \quad d = 3,$$

$K \subset G_2 \cap \mathfrak{K}_3$, $G_2 \in \mathcal{G}_2^{(3)}$ (the case $k = 1 = 2$). These relations are easily verified using (2.2.10)-(2.2.11), (2.2.13)-(2.2.14), and Hadwiger's theorem.

Let X_d be a d-polytope and

$$\mathcal{F}_k(X_d) = \{\text{k-faces of } X_d\} .$$

Recall that a bounded Voronoi cell is a d-polytope comprising a hierarchy of faces in its boundary, cf. Remark 2.1.2. The foregoing discussion suggests the following quantities as natural characteristics of $X_k \in \mathcal{F}_k(X_d)$, $1 \leq k \leq d$:

$d = 3$: $V(X_3)$, $S(X_3)$, $\bar{b}(X_3)$, $\bar{L}(X_3)$, $N_2(X_3)$, $N_1(X_3)$, $N_0(X_3)$

$$(2.2.16)$$

$d = 2,3$: $A(X_2)$, $U(X_2)$, $N(X_2)$, $L(X_1)$. $\qquad (2.2.17)$

Here

$$\bar{L}(X_3) = \sum_{F_1 \in \mathcal{F}_1(X_3)} L(F_1)$$

is the total length of all edges contained in the polyhedron X_3, and $N_0(X_3)$, $N_1(X_3)$, $N_2(X_3)$ are the number of vertices, edges, and sides of the polyhedron X_3, respectively, while $N(X_2) = N_0(X_2) = N_1(X_2)$ is the number of vertices (or equivalent edges) of the polygon X_2.

All these characteristics, except the mean breadth, have simple interpretations as number, length, area and volume. The mean breadth is connected to the edges and angles of X_3 as follows (see e.g. Santaló, 1976):

$$\bar{b}(X_3) = \frac{1}{4\pi} \sum_{F_1 \in \mathcal{F}_1(X_3)} \gamma(F_1,X_3)L(F_1) \qquad (2.2.18)$$

where $\gamma(F_1,X_3)$ is the exterior angle of the edge F_1 with respect to X_3, i.e. the angle between the normals of the two halfspaces which bound X_3 and contains F_1. This is related to the associated interior angle $\phi(F_1,X_3)$ by

$$\gamma(F_1,X_3) = \pi-\phi(F_1,X_3) \qquad (2.2.19)$$

see Figure 2.2.2.

<u>Figure 2.2.2</u>. Interior angle ϕ and exterior angle γ at a vertex $(d = 2)$ or at an edge $(d = 3)$; in the latter case the figure should be considered as a plane section perpendicular to the edge. It follows that $\gamma+\phi = 2\pi-\pi = \pi$.

Recall also Euler's relation,

$$N_0(X_3) - N_1(X_3) + N_2(X_3) = 2 \ , \quad d = 3 \ . \tag{2.2.20}$$

There is an analogous relation concerning the interior angles, called Gram's relation, which states that

$$\sum_{k=0}^{d} (-1)^k \sum_{F_k \in \mathscr{F}_k(X_d)} \phi(F_k, X_d) = 0 \tag{2.2.21}$$

cf. Grünbaum (1967, Ch. 14.1). Here $\phi(X_d, X_d) := 2c_d$ and the interior angle of a vertex $F_0 = \{x\} \in \mathscr{F}_0(X_d)$ is given by

$$\phi(F_0, X_d) = \lim_{\epsilon \to 0} v_d(\{y \in X_d \mid \|y - x\| = \epsilon\}) / \epsilon^{d-1} \tag{2.2.22}$$

For $F_{d-1} \in \mathscr{F}_{d-1}(X_d)$ we have simply that $\phi(F_{d-1}, X_d) = c_d$.

In the remaining part of this section we state some propositions which have important stereological applications. Stereology concerns the determination of internal properties of a geometric object from lower-dimensional information such as that provided by linear or planar sections through the object.

Proposition 2.2.1 (Crofton's section formulae). We have that

$$A(K) = \int L(K \cap G_1) dG_1 \ , \qquad K \in \mathcal{K}_2 \ , \tag{2.2.23}$$

$$U(K) = \pi \int \chi(K \cap G_1) dG_1 \ , \qquad K \in \mathcal{K}_2 \ , \tag{2.2.24}$$

$$V(K) = \int L(K \cap G_1) dG_1 \ , \qquad K \in \mathcal{K}_3 \ , \tag{2.2.25}$$

$$S(K) = 4 \int \chi(K \cap G_1) dG_1 \ , \qquad K \in \mathcal{K}_3 \ , \tag{2.2.26}$$

$$V(K) = \int A(K \cap G_2) \, dG_2 , \qquad K \in \mathcal{K}_3 , \qquad (2.2.27)$$

$$S(K) = \frac{4}{\pi} \int U(K \cap G_2) \, dG_2 , \qquad K \in \mathcal{K}_3 , \qquad (2.2.28)$$

$$\bar{b}(K) = \int \chi(K \cap G_2) \, dG_2 , \qquad K \in \mathcal{K}_3 . \qquad (2.2.29)$$

$\square$

Proof. Observe that (2.2.23), (2.2.25), and (2.2.27) follow from the following more general results:

$$A(K) = \int_{L_1^{\perp}} L(K \cap (L_1 + y^1)) \, dy^1 , \qquad d = 2 , \qquad (2.2.30)$$

if $L_1 \in \mathcal{L}_1^{(2)}$ is fixed,

$$V(K) = \int_{L_1^{\perp}} L(K \cap (L_1 + y^2)) \, dy^2 , \qquad d = 3 , \qquad (2.2.31)$$

if $L_1 \in \mathcal{L}_1^{(3)}$ is fixed, and

$$V(K) = \int_{L_2^{\perp}} A(K \cap (L_2 + y^1)) \, dy^1 , \qquad d = 3 , \qquad (2.2.32)$$

if $L_2 \in \mathcal{L}_2^{(3)}$ is fixed. Formulae (2.2.30)-(2.2.32) hold for any Borel set $K \subset \mathbb{R}^d$ and they are all easy consequences of Fubini's theorem. (Formulae (2.2.30) and (2.2.32) are the so-called avalierie principle). Further, (2.2.29) is just (2.2.13) combined with the definition (2.2.7). Formulae (2.2.24) and (2.2.28) can be verified either using Hadwiger's theorem or by direct calculations when K is a polytope. Finally, combining (2.2.1), (2.2.2), (2.2.24), and (2.2.28) we obtain (2.2.26).

$\square$

<u>Proposition 2.2.2</u> (Crofton's chord formula). For $K \in \mathcal{K}_d$,

$$\frac{\pi}{3} \int L(K \cap G_1)^{d+1} \, dG_1 = \begin{cases} A(K)^2 , & d = 2 \\ V(K)^2 , & d = 3 \end{cases} \qquad (2.2.33)$$

Especially, if $d = 3$ and K is contained in a 2-flat,

$$\frac{4}{3} \int L(K \cap G_2)^3 \, dG_2 = A(K)^2 . \qquad \qquad \square$$

<u>Proof</u>. See Exercise 2.2.3. $\qquad \qquad \square$

Finally, we shall establish an extremely useful decomposition of the Lebesgue measure $dx_0 \cdots dx_k$ with respect to the k-flat $G_k = z + L_k$ as defined by Lemma 2.1.1 when $1 \leq k \leq d$ (more precisely, z and L_k are well-defined except on a Lebesgue nullset). Let R be the radius of the sphere in G_k through $x_0, \ldots, x_k$, and let $u_i \in L_k \cap S^{d-1}$ be the direction of z to x_i, that is

$$x_i = z + R u_i , \quad i = 0, \ldots, k , \qquad (2.2.34)$$

see Figure 2.2.3. Furthermore, let $v = v(u_0, \ldots, u_k)$ be $k!$ times the k-content of the (k+1)-simplex in L_k with vertices $u_0, \ldots, u_k$.

<u>Proposition 2.2.3</u> (Blaschke-Petkantschin's formula). We have

$$dx_0 \cdots dx_k = c_k'^{(d)} \, v^{d-k+1} \, R^{dk-1} \, dR \, v_k(du_0) \cdots v_k(du_k) dz dL_k \qquad (2.2.35)$$

where v_k denotes surface measure on the unit sphere $L_k \cap$ S^{d-1} and $\acute{c}_1^{(2)} = \pi$, $\acute{c}_1^{(3)} = \acute{c}_2^{(3)} = 2\pi$, $\acute{c}_d^{(d)} = 1$. □

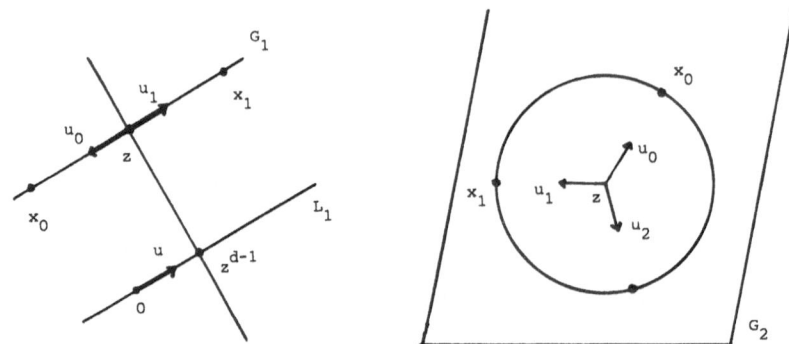

<u>Figure 2.2.3</u>. Left: $k = 1$. Right: $k = 2$.

<u>Proof</u>. Let us verify the cases of (2.2.35) when $k = 1$ or $d = k$; the proof of the remaining case $d = 3$ and $k = 2$ is similar but somewhat more technical (general proofs can be found in e.g. Miles (1974) and Møller (1985)).

Consider Figure 2.2.3 when $k = 1$. We have two situations: $u_0 = -u_1 = u$ or $u_0 = -u_1 = -u$. Note that v^d $v_1(du_0)v_1(du_1) = 2^d$ in both cases and equals 0 otherwise. Suppose for instance that $u_0 = -u_1 = -u$ (as in Figure 2.2.3). Then the one-to-one mapping

$$(z,R,u) \rightarrow (x_0,x_1) = (z-Ru, z+Ru)$$

has Jacobian

$$J = \left| \begin{array}{c} \dfrac{\partial x_0}{\partial(z,R,u)}^* \\[2ex] \dfrac{\partial x_1}{\partial(z,R,u)}^* \end{array} \right| = \left| \begin{array}{ccc} I_d & -u & -R\dot{u} \\[2ex] I_d & u & R\dot{u} \end{array} \right|$$

Here $\partial u = \partial\theta$ and $\dot{u} = \partial u/\partial\theta$ if $d = 2$, whereas $\partial u = \partial(\theta,\phi)^*$ and $\dot{u} = [\partial u/\partial\theta \ \ \partial u/\partial\phi]$ if $d = 3$, where we have used a parametrization as in (2.2.3) and (2.2.4), respectively. Thus

$$J = R^{d-1} \left\| \begin{bmatrix} I_d & I_d \\ -u^* & u^* \\ -\dot{u}^* & \dot{u}^* \end{bmatrix} \begin{bmatrix} I_d & -u & -\dot{u} \\ I_d & u & \dot{u} \end{bmatrix} \right\|^{1/2}$$

$$= R^{d-1} \left| \begin{matrix} 2I_d & 0 \\ 0 & 2I_d \end{matrix} \right|^{1/2} = 2^d R^{d-1}$$

where we have used that $[u \ \dot{u}] \in O(d)$ so

$$u^* u = 1, \quad \dot{u}^* u = 0, \quad \dot{u}^* \dot{u} = I_{d-1}. \qquad (2.2.36)$$

Hence

$$dx_0 dx_1 = v^d R^{d-1} dR \ v_1(du_0) v_1(du_1) v_d(du)$$

whereby (2.2.35) follows from (2.2.1).

Next, let $d = k$. Then the Jacobian of the injective mapping $(z, R, u_0, \ldots, u_d) \to (x_0, \ldots, x_d)$ becomes

$$J = \left| \begin{matrix} I_d & u_0 & R\dot{u}_0 & \cdots & 0 \\ \vdots & \vdots & \vdots & \ddots & \vdots \\ I_d & u_d & 0 & \cdots & R\dot{u}_d \end{matrix} \right|,$$

so using (2.2.36) with u replaced by u_i, $i = 0, \ldots, d$,

$$J = R^{(d+1)(d-1)} \left\| \begin{bmatrix} I_d & \cdots & I_d \\ u_0^* & \cdots & u_d^* \\ \dot{u}_0^* & \cdots & 0 \\ \vdots & \ddots & \vdots \\ 0 & \cdots & \dot{u}_d^* \end{bmatrix} \begin{bmatrix} I_d & u_0 & \dot{u}_0 & \cdots & 0 \\ \vdots & \vdots & \vdots & \ddots & \vdots \\ I_d & u_d & 0 & \cdots & \dot{u}_d \end{bmatrix} \right\|^{1/2}$$

$$= R^{d^2-1} \begin{vmatrix} (d+1)I_d & \sum\limits_0^d u_i & \dot{u}_0 & \cdots & \dot{u}_d \\ \sum\limits_0^d u_i^* & d+1 & 0 & \cdots & 0 \\ \dot{u}_0^* & 0 & I_{d-1} & \cdots & 0 \\ \vdots & \vdots & \vdots & \ddots & \vdots \\ \dot{u}_d^* & 0 & 0 & \cdots & I_{d-1} \end{vmatrix}^{1/2}$$

Now, use the well-known formula $|\Sigma| = |\Sigma_{11 \cdot 2}||\Sigma_{22}|$ where

$$\Sigma = \begin{bmatrix} \Sigma_{11} & \Sigma_{12} \\ \Sigma_{21} & \Sigma_{22} \end{bmatrix}, \quad \Sigma_{11 \cdot 2} = \Sigma_{11} - \Sigma_{12} \Sigma_{22}^{-1} \Sigma_{21},$$

with Σ a symmetric matrix so that and Σ_{22}^{-1} exists, to obtain that

$$J = R^{d^2-1} |\Sigma_{11 \cdot 2}|^{1/2} \qquad (2.2.37)$$

with $\Sigma_{22} = I_{(d+1)(d-1)}$ and

$$\Sigma_{11 \cdot 2} = \begin{bmatrix} (d+1)I_d - \sum\limits_0^d \dot{u}_i \dot{u}_i^* & \sum\limits_0^d u_i \\ \sum\limits_0^d u_i^* & d+1 \end{bmatrix}.$$

Since $[u_i \; \dot{u}_i]$ is an orthogonal $d \times d$ matrix we have that

$$I_d - \dot{u}_i \dot{u}_i^* = [u_i \; \dot{u}_i][u_i \; \dot{u}_i]^* - \dot{u}_i \dot{u}_i^* = u_i u_i^*$$

and so

$$\Sigma_{11 \cdot 2} = \sum_0^d \begin{bmatrix} u_i u_i^* & u_i \\ u_i^* & 1 \end{bmatrix} = \sum_0^d \begin{bmatrix} u_i \\ 1 \end{bmatrix} [u_i^* \; 1] \; .$$

Hence,

$$|\Sigma_{11 \cdot 2}|^{1/2} = \left| \sum_0^d \begin{bmatrix} u_i \\ 1 \end{bmatrix} [u_i^* \; 1] \right|^{1/2} = \left| \begin{matrix} u_0 & & u_d \\ 1 & \cdots & 1 \end{matrix} \right| = \triangledown \; .$$

Inserting this in (2.2.37) gives (2.2.35). □

Exercise 2.2.1. Show that the measure dG_k is invariant under motions in $\mathbb{R}^d$ when $1 \leq k < d = 2, 3$.

Exercise 2.2.2. Verify formulae (2.2.11) and (2.2.14) by using Hadwiger's theorem.

Exercise 2.2.3. Prove Proposition 2.2.2.

Hint: The Blaschke-Petkantshin formula (2.2.35) can be used.

Exercise 2.2.4. Let $K \in \mathcal{K}_d$ such that

$$s_1^{(d)} = \int \chi(K \cap G_1) dG_1 > 0 \quad \text{when} \quad d = 2, 3 \; ,$$

and

$$s_2^{(3)} = \int \chi(K \cap G_2) dG_2 > 0 \quad \text{when} \quad d = 3 \; .$$

Define the following probability measures on $\mathcal{G}_k^{(d)}$:

$$\mu_1^{(d)}(dG_1) = \chi(K \cap G_1)dG_1/s_1^{(d)} , \quad d = 2,3 , \quad k = 1 ,$$

$$\mu_2^{(3)}(dG_2) = \chi(K \cap G_2)dG_2/s_2^{(3)} , \quad d = 3 , \quad k = 2 .$$

A random line with distribution $\mu_1^{(d)}$ is called an IUR-line through K, while a random plane with distribution $\mu_2^{(3)}$ is called an IUR-plane through K (IUR means 'isotropic uniform random'). Let $\bar{G}_1$ be an IUR-line through K ($d = 2,3$) and $\bar{G}_2$ be an IUR-plane through K ($d = 3$). Show that

$$EL(K \cap \bar{G}_1) = \pi A(K)/U(K) , \quad d = 2 , \qquad (2.2.38)$$

$$EL(K \cap \bar{G}_1) = 4V(K)/S(K) , \quad d = 3 , \qquad (2.2.39)$$

$$EU(K \cap \bar{G}_2) = \frac{\pi}{4} S(K)/\bar{b}(K) , \quad d = 3 , \qquad (2.2.40)$$

$$EA(K \cap \bar{G}_2) = V(K)/\bar{b}(K) , \quad d = 3 , \qquad (2.2.41)$$

$$E[L(K \cap \bar{G}_1)^3] = 3A(K)^2/U(K) , \quad d = 2 , \qquad (2.2.42)$$

$$E[L(K \cap \bar{G}_1)^4] = \frac{12}{\pi} V(K)^2/S(K) , \quad d = 3 . \qquad (2.2.43)$$

In addition show that for a polyhedron K in $\mathbb{R}^3$,

$$EN(K \cap \bar{G}_2) = \frac{1}{2} \bar{L}(K)/\bar{b}(K) , \qquad (2.2.44)$$

while if K is contained in a 2-flat in $\mathbb{R}^3$,

$$EL(K \cap \bar{G}_2)^3 = 3A(K)^2/U(K) \qquad (2.2.45)$$

and especially if K is a polygon in $\mathbb{R}^3$,

$$EL(K \cap \bar{G}_2) = \pi A(K)/U(K) \ . \qquad\qquad (2.2.46)$$

Finally, discuss how formulae (2.2.38)-(2.2.46) can be used to suggest unbiased stereological estimators of ratios between certain characteristics of a bounded Voronoi cell and its lower-dimensional faces.

We return to this exercise in Exercise 3.4.2.

3. Stationary Voronoi tessellations

A random tessellation is said to be stationary if its distribution is invariant under translations in $\mathbb{R}^d$. Assuming stationarity it is possible to define what is meant by a typical cell and a typical k-facet of the tessellation. The objective in this chapter is to formalize these concepts and study their relationships for stationary Voronoi and Delaunay tessellations. In fact all results presented for stationary Voronoi tessellations hold as well for arbitrary normal stationary tessellations with convex cells. 'Normality' in this context means that every k-facet lies in the boundaries of exactly d-k+1 cells, k = 0,...,d-1. Many real-life non-artificial tessellations for d = 1,2,3 possess this property. Indeed Voronoi tessellations in general quadratic position are normal, cf. Proposition 2.1.1. Though Delaunay tessellations are not, many results for Voronoi tessellations carry over because of the duality.

Section 3.1 contains a brief account on spatial point processes which is needed for the following Sections 3.2-3.4. The approach in Sections 3.2-3.4 on Palm distributions is much inspired by the so-called East German school, see e.g. Stoyan et al. (1987) and Mecke et al. (1990). More general results, including those for non-normal tessellations, can be found in these monographs and in papers by Zähle (1988) and Møller (1989). In Section 3.2 we study the Palm distributions of typical Voronoi and Delaunay cells and facets. In Section 3.3 some mean value relations between moments of certain cell and facet characteristics are established. Finally, in Section 3.4 we establish further relations between characteristics of a Voronoi tessellation and its intersection with an affine subspace. Again many of these results extend to more general types of stationary tessellations, cf. e.g. Møller (1989).

3.1. Spatial point processes and stationarity

The distributions of the Voronoi tessellation $\mathcal{V}(\Phi)$ and the Delaunay tessellation $\mathcal{T}(\Phi)$ are of course determined by the distribution of the nuclei Φ. This can be described as a spatial point process since Φ is assumed to be locally finite. In this section we present some background material on general spatial point processes; more details can be found in e.g. Stoyan et al. (1987) and Daley and Vere-Jones (1988).

Let Ω denote the space of all subsets $\varphi \subset \mathbb{R}^d$ which are locally finite. For any bounded Borel set $B \subset \mathbb{R}^d$, define N_B: $\Omega \rightarrow \{0,1,\ldots\}$ by $N_B(\varphi) = \#(\varphi \cap B)$, the number of points contained in $\varphi \cap B$. Let $\mathcal{F}$ be the smallest σ-field making the mappings N_B measurable for all bounded Borel sets $B \subset \mathbb{R}^d$. Then any measurable mapping Φ defined on some probability space and taking values in Ω is called a spatial point process on $\mathbb{R}^d$. (Strictly speaking, this is the definition of a simple point process, i.e. when Φ has no multiple points).

Let Φ be a spatial point process on $\mathbb{R}^d$. It can be shown that $\mathcal{F}$ is also the σ-field generated by the sets $F_K = \{\varphi \in \Omega \mid \varphi \cap K \neq \emptyset\}$, $K \in \mathcal{K}_d$ (the compact subsets of $\mathbb{R}^d$, cf. Section 2.2). Thus the distribution of Φ is uniquely determined by the void-probabilities,

$$v_K = P(\Phi \cap K = \emptyset), \quad K \in \mathcal{K}_d . \qquad (3.1.1)$$

The point process $\Phi = \{x_i\}$ is said to be stationary if its distribution is invariant under translations in $\mathbb{R}^d$. This means that Φ is distributed as

$$\Phi + y = \{x_i + y \mid x_i \in \Phi\}$$

for all $y \in \mathbb{R}^d$. Similarly, Φ is said to be isotropic if its distribution is invariant under rotations about the origin in $\mathbb{R}^d$, i.e. when Φ is distributed as

$$O\Phi = \{Ox_i \mid x_i \in \Phi\}$$

for all $O \in O(d)$. Here $O(d)$ denotes the orthogonal group of $d \times d$ matrices, i.e., $O \in O(d)$ if and only if $O^{-1} = O^*$.

Suppose Φ is stationary. Then

$$\Lambda(B) = EN_B(\Phi) \tag{3.1.2}$$

is seen to be a translation invariant measure on the Borel σ-field $\mathscr{B}$ in $\mathbb{R}^d$, and so

$$\Lambda(B) = \lambda |B|_d \tag{3.1.3}$$

for some constant $\lambda \in [0,\infty]$ called the intensity of Φ. Thus λ is the mean number of points in Φ per unit volume. Clearly, $\lambda = 0$ if and only if $\Phi = \emptyset$ almost surely. Assume that $0 < \lambda < \infty$ and consider the measure

$$\mu(B\times F) = E \sum_{x_i \in \Phi \cap B} 1(\Phi - x_i \in F) \tag{3.1.4}$$

defined on the product space $(\mathbb{R}^d \times \Omega, \mathscr{B} \otimes \mathscr{F})$, where $1(\cdot)$ is the indicator function. Then, for any $y \in \mathbb{R}^d$ we have that

$$\mu((B+y)\times F) = E \sum_{x_i \in \Phi \cap (B+y)} 1(\Phi - x_i \in F)$$

$$= E \sum_{x_i - y \in (\Phi - y) \cap B} 1((\Phi - y) - (x_i - y) \in F)$$

$$= E \sum_{x_i \in \Phi \cap B} 1(\Phi - x_i \in F)$$

$$= \mu(B \times F)$$

because of the stationarity. Therefore, $B \to \mu(B \times F)$ is a translation invariant measure on $\mathcal{B}$, so

$$\mu(B \times F) = |B|_d \, q(F) \tag{3.1.5}$$

for some non-negative function $q(F)$. Letting $0 < |B|_d < \infty$ then (3.1.4) and (3.1.5) imply that

$$q(F) = E \sum_{x_i \in \Phi \cap B} 1(\Phi - x_i \in F) / |B|_d \tag{3.1.6}$$

is a measure on $(\Omega, \mathcal{F})$ which does not depend on the specific choice of B. Combining (3.1.2)-(3.1.3) and (3.1.6) give that

$$q(\Omega) = E N_B(\Phi) / |B|_d = \lambda \ ,$$

so

$$Q(F) = E \sum_{x_i \in \Phi \cap B} 1(\Phi - x_i \in F) / (\lambda |B|_d) \tag{3.1.7}$$

becomes a probability measure on $(\Omega, \mathcal{F})$ called the Palm distribution of Φ. Exercise 3.1.1 demonstrates that this can be given an interpretation as the conditional distribution of $\Phi - y$ given that Φ has a point situated at an arbitrary location $y \in \mathbb{R}^d$. Note that the Palm distribution is not necessarily stationary. However, isotropy of Φ implies that the Palm distribution becomes isotropic too. Finally, let us notice that (3.1.7) and standard measure theoretical techniques establish a socalled Campbell theorem, which states that

$$\lambda \int E_Q f(\Phi, y) dy = E \sum_{x_i \in \Phi} f(\Phi - x_i, x_i) \tag{3.1.8}$$

where $f: \Omega \times \mathbb{R}^d \to [0,\infty)$ is an arbitrary measurable function and E_Q denotes expectation with respect to the Palm distribution.

In this chapter we assume stationarity. This implies that the statistical properties of $\mathscr{V}(\Phi)$ and $\mathscr{T}(\Phi)$ do not depend on the location of the observer. Of course, in many applications this may seem as being a strict assumption, but often it must be assumed in order to make progress in any analysis. At least within sufficient small 'homogeneous' regions it may be reasonable to assume stationarity.

The next proposition shows that stationarity implies that all Voronoi cells are bounded and the Delaunay tessellation becomes a tessellation of the entire space $\mathbb{R}^d$.

Proposition 3.1.1. The condition (c) in Remark 2.1.3 holds almost surely when Φ is stationary and $\Phi \neq \emptyset$ almost surely.

□

Proof. For $y = (y_1,\ldots,y_d) \in \mathbb{Z}^d$ with $y_i \neq 0$, $i = 1,\ldots,d$, define the unbounded 'rectangle'

$$R(y) = \{x=(x_1,\ldots,x_d)\in\mathbb{R}^d | x_i > \text{sign}(y_i)y_i, \ i=1,\ldots,d\}$$

where $\text{sign}(y_i)$ denotes the sign of y_i. Then (c) holds with probability one if

$$P(\Phi \cap R(y)=\emptyset) = 0$$

for all such points y. Suppose this is not the case for some y. We can without loss of generality assume that $y_i > 0$, $i = 1,\ldots,d$. Then

$$P(\Phi=\emptyset)$$

$$= \lim_{h\to-\infty} P(\Phi\cap\{(x_1,\ldots,x_d)\in\mathbb{R}^d|x_i>h,\ i=1,\ldots,d\}=\emptyset)$$

$$= \lim_{h\to-\infty} P(\{\Phi-(h-y_1,\ldots,h-y_d)\}\cap R(y)=\emptyset)$$

$$= P(\Phi\cap R(y)=\emptyset)$$

by the stationarity. Hence $\Phi = \emptyset$ with a positive probability. This contradiction proves the assertion. □

Exercise 3.1.1. Let $B = \{x\in\mathbb{R}^d|\|x-y\|<r\}$ be the ball centered at an arbitrary point $y \in \mathbb{R}^d$ and with radius $r > 0$. Demonstrate by heuristics that the Palm measure given by (3.1.7) may be expected to satisfy that

$$\lambda|B|_d Q(F) \cong P(\Phi-y\in F|N_B(\Phi)=1)P(N_B(\Phi)=1)$$

$$\cong P(\Phi-y\in F|y\in\Phi)\lambda|B|_d$$

as $r \to 0$, and so

$$Q(F) \cong P(\Phi-y\in F|y\in\Phi) .$$

(Strictly speaking, the latter conditional probability makes no formal sense as $P(y\in\Phi) = 0$ because of the stationarity).

3.2. Palm measures and intensities of cells and facets

Throughout this section Φ denotes a stationary point process on $\mathbb{R}^d$ with an intensity $\lambda \in (0,\infty)$ as defined in the previous section. Moreover, Φ is assumed to be in general quadratic position with probability one. Then according to Proposition 3.1.1 and the discussion in Sections 1.1 and 2.1,

$\mathcal{V}(\Phi)$ and $\mathcal{T}(\Phi)$ become both tessellations of $\mathbb{R}^d$ with in-finite but countable many cells. Moreover, the Voronoi cells are d-polytopes and the Delaunay cells are d-dimensional sim-plicies. Recall also that for both tessellations, the k-facets agree with the k-faces of their cells, $k = 0, \ldots, d$.

The objective in this section is to define and describe the distributions of the 'typical' k-facets of $\mathcal{V}(\Phi)$ and $\mathcal{T}(\Phi)$, using Palm measure techniques. For this we need to in-troduce various point processes associated with the sets of facets. Finally, we shall establish some topological mean value relations between the intensities of these point processes.

Let us for simplicity first specify what is meant by the 'typical' Voronoi cell. To each nucleus $x_i \in \Phi$ we associate a point $z(x_i | \Phi) \in \mathbb{R}^d$ which is equivariant under translations in $\mathbb{R}^d$, that is

$$z(x_i + y | \Phi + y) = y + z(x_i | \Phi) \qquad (3.2.1)$$

for all $y \in \mathbb{R}^d$. We call $z(x_i | \Phi)$ the centroid of the cell $C(x_i | \Phi)$. Example of centroids are

(i) $z(x_i | \Phi) = x_i$, the nucleus of the cell,

(ii) $z(x_i | \Phi) =$ center of gravity of $C(x_i | \Phi)$, i.e. the arithmetic average of the vertices of $C(x_i | \Phi)$,

(iii) $z(x_i | \Phi) =$ the 'extreme' vertex of $C(x_i | \Phi)$ with respect to a given direction.

The discussion and results presented in the following do not depend on the specific choice of centroids as argued in Remark 3.2.1 below. Sometimes (i) is most convenient to use for proof technical reasons, while (ii) may be natural for physical reasons, whereas (iii) may be easier to use for sampling purposes. In each of the cases (i)-(ii), but not in the case (iii), the collection of centroids parametrizes the collection of Voronoi cells. Note also that the centroids in (ii) and (iii) are functions $z(x_i|\Phi) = z(C(x_i|\Phi))$ of the cell only, while this is indeed not the case in (i).

Now, we proceed as in Section 3.1 (for details, see Exercise 3.2.1). Using (2.2.1) and the stationarity of Φ, we find that

$$\lambda_d = E \sum_{x_i \in \Phi} 1(z(x_i|\Phi) \in B) / |B|_d \qquad (3.2.2)$$

does not depend on $B \in \mathcal{B}$ with $0 < |B|_d < \infty$. The constant λ_d is called the intensity of Voronoi cells. Assume that $0 < \lambda_d < \infty$ (this is in fact the case, cf. Remark 3.2.1). Let Γ be the set of polytopes in $\mathbb{R}^d$ equipped with a suitable σ-field $\mathcal{A}$ (by 'suitable' we mean of course that all mappings considered in the following should be measurable). Then the Palm measure of the typical Voronoi cell is the distribution on $(\Gamma, \mathcal{A})$ defined by

$$Q_d(A) = E \sum_{x_i \in \Phi} 1(z(x_i|\Phi) \in B, \ C(x_i|\Phi) - z(x_i|\Phi) \in A) / (\lambda_d |B|_d)$$

$$(3.2.3)$$

for an arbitrary set $B \in \mathcal{B}$ with $0 < |B|_d < \infty$ (the right hand side of (3.2.3) does not depend on B). The typical Voronoi cell is defined as the random polytope $\mathcal{C}_d$ on $(\Gamma, \mathcal{A})$ having distribution Q_d. This can be interpreted as the conditional distribution of $C(x_0|\Phi) - y$ given that Φ has a nucleus x_0 with centroid $z(x_0|\Phi) = y$ where $y \in \mathbb{R}^d$ is an arbitrary point; an alternative interpretation is given in Remark 3.3.2 in the next section. Especially, if the nuclei are used as centroids (the case (i) above), $\mathscr{C}_d$ is distributed as $C(0|\overset{\prime}{\Phi})$ where the distribution of the point process $\overset{\prime}{\Phi}$ is given by the Palm distribution of Φ as defined in Section 3.1. From (3.2.3) we obtain another Campbell theorem:

$$\lambda_d \int Ef(\mathscr{C}_d, y)\,dy = E \sum_{x_i \in \Phi} f(C(x_i|\Phi) - z(x_i|\Phi), z(x_i|\Phi)) \quad (3.2.4)$$

where f is any non-negative measurable function.

Next, let us consider the general case of Voronoi k-facets, $0 \leq k \leq d$. Let Φ_k be the set of all point configurations $x = \{x_0, \ldots, x_{d-k}\} \subset \Phi$ of $d-k+1$ distinct nuclei with $F(x|\Phi) = F(x_0, \ldots, x_{d-k}|\Phi) \neq \emptyset$. Especially, $\Phi_d = \Phi$. To each point configuration $x = \{x_0, \ldots, x_{d-k}\} \in \Phi_k$ we associate an equivariant centroid $z(x|\Phi) \in \mathbb{R}^d$ such that

$$z(x+y|\Phi+y) = y + z(x|\Phi)$$

for all $y \in \mathbb{R}^d$, where $x+y = \{x_0+y, \ldots, x_{d-k}+y\}$. There are several possibilities for choosing $z(x|\Phi)$, examples include

(i) $z(x|\Phi) = z(x_0, \ldots, x_{d-k})$ as defined in Lemma 2.1 (if $k = d$ we take $z(x_0|\Phi) = x_0$),

(ii)   the center of gravity of the k-facet $F(x|\Phi)$,

(iii)   the most 'extreme' vertex of $F(x|\Phi)$ in some given direction.

Here (i) turns out to be convenient for some proofs presented later on, while (iii) might be convenient for the purpose of sampling facets in practice. But as remarked before the specific choice of centroid is actually not important for the results presented in these notes.

The intensity of k-facets is well-defined by

$$\lambda_k = E \sum_{x \in \Phi_k} 1(z(x|\Phi) \in B) / |B|_d \qquad (3.2.5)$$

for an arbitrary set $B \in \mathcal{B}$ with $0 < |B|_d < \infty$. Clearly, $\lambda_k > 0$. Assuming that $\lambda_k < \infty$, the Palm distribution of the typical Voronoi k-facet is well-defined by

$$Q_k(A) = E \sum_{x \in \Phi_k} 1(z(x|\Phi) \in B, F(x|\Phi) - z(x|\Phi) \in A) / (\lambda_k |B|_d) \qquad (3.2.6)$$

with $A \in \mathcal{A}$, $B \in \mathcal{B}$, $0 < |B|_d < \infty$. We denote by $\mathcal{C}_k$ the random polytope on $(\Gamma, \mathcal{A})$ with distribution Q_k. Roughly speaking, $\mathcal{C}_k$ is distributed as $F(x|\Phi) - y$ given that Φ_k contains a configuration of $d-k+1$ nuclei with centroid $z(x|\Phi) = y$ at an arbitrary location $y \in \mathbb{R}^d$. Moreover, we have a Campbell theorem:

$$\lambda_k \int Ef(\mathcal{C}_k, y) dy = E \sum_{x \in \Phi_k} f(F(x|\Phi) - z(x|\Phi), z(x|\Phi)) \qquad (3.2.7)$$

for non-negative measurable functions f. In addition, if the centroid is equivariant under rotations, i.e. $z(\mathcal{O}x|\mathcal{O}\Phi) = \mathcal{O}z(x|\Phi)$ for all $\mathcal{O} \in \mathcal{O}(d)$ and $x \in \Phi_k$, then isotropy of Φ implies isotropy of $\mathcal{C}_k$, see Exercise 3.2.2. The examples (i)-(iii) of centroids are all equivariant under rotations.

Remark 3.2.1. In the following we shall only consider statements which involve quantities like λ_k and $Ef(\mathcal{C}_k)$ where $f: \Gamma \to [0, \infty)$ is measurable and invariant under translations in $\mathbb{R}^d$, i.e., $f(F+y) = f(F)$ for all $y \in \mathbb{R}^d$ and $F \in \Gamma$. Then λ_k as well as $Ef(\mathcal{C}_k)$ do not depend on the specific choice of the centroid. This follows from the next lemma.

Consequently, $\lambda_d = \lambda$ and λ_0 is the mean number of vertices of $\mathscr{V}(\Phi)$ per unit volume. □

Lemma 3.2.1. For $0 \leq k \leq d$ let $B \in \mathscr{B}$ with $0 < |B|_d < \infty$ if $k < d$ and $0 \leq |B|_d < \infty$ if $k = d$. Then

$$\lambda_k \, \text{Ef}(\mathscr{C}_k) = \text{E} \sum_{x \in \Phi_k} 1[0 \in F(x|\Phi) \oplus B] f(F(x|\Phi)) / |F(x|\Phi) \oplus B|_d \quad (3.2.8)$$

if f is a non-negative measurable function which is invariant under translations. Here $\oplus$ denotes Minkowski sum of subsets of $\mathbb{R}^d$, i.e.

$$B_1 \oplus B_2 = \{b_1 + b_2 | b_1 \in B_1, b_2 \in B_2\} \ . \qquad □$$

Proof. According to (3.2.7) we have that

$$\lambda_k \, \text{Ef}(\mathscr{C}_k)$$

$$= \lambda_k \int \text{E} \ 1(-y \in \mathscr{C}_k \oplus B) \ \frac{f(\mathscr{C}_k)}{|\mathscr{C}_k \oplus B|_d} \ dy$$

$$= \text{E} \sum_{x \in \Phi_k} 1(-z(x|\Phi) \in (F(x|\Phi) - z(x|\Phi)) \oplus B) \ \frac{f(F(x|\Phi) - z(x|\Phi))}{|(F(x|\Phi) - z(x|\Phi)) \oplus B|_d}$$

$$= \text{E} \sum_{x \in \Phi_k} 1(0 \in F(x|\Phi) \oplus B) \ \frac{f(F(x|\Phi))}{|F(x|\Phi) \oplus B|_d}$$

as f is invariant under translations. □

The following propositions relate λ_k and $\mathscr{C}_k$ to λ and $\mathscr{C}_d$. The propositions are commented in Remark 3.2.2 below. Recall that for a d-polytope X_d, $\mathscr{F}_d(X_d)$ denotes the set of k-faces contained in X_d.

Proposition 3.2.1. Let $N_k(\mathscr{C}_d)$ be the number of k-faces contained in $\mathscr{C}_d$. Then

$$\lambda_k = \lambda E N_k(\mathscr{C}_d)/(d-k+1) , \quad k = 0,\ldots,d . \qquad (3.2.9)$$

□

Proof. Since N_k is invariant under translations in $\mathbb{R}^d$ we can choose the centroid $z(x|\Phi)$ as a function of $F(x|\Phi)$ only, so $z(x|\Phi) = z(F(x|\Phi))$, say. Then the equivariance of the centroid implies that

$$|B|_d \, \lambda E N_k(\mathscr{C}_d) = \lambda \int E \sum_{F_k \in \mathscr{F}_k(\mathscr{C}_d)} 1(z(y+F_k) \in B) \, dy$$

$$= \lambda \int E \sum_{F_k \in \mathscr{F}_k(\mathscr{C}_d+y)} 1(z(F_k) \in B) \, dy$$

and so by (3.2.4),

$$= E \sum_{x_i \in \Phi} \sum_{F_k \in \mathscr{F}_k(C(x_i|\Phi))} 1(z(F_k) \in B)$$

$$= E \sum_{x_i \in \Phi} \sum_{x \in \Phi_k} 1(z(x|\Phi) \in B, \, x_i \in x)$$

$$= E \sum_{x \in \Phi_k} (d-k+1) 1(z(x|\Phi) \in B)$$

$$= |B|_d (d-k+1) \lambda_k$$

cf. (3.2.5). This proves (3.2.9). □

<u>Proposition 3.2.2</u>. The distribution of $\mathscr{C}_k$ is related to $\mathscr{C}_d$ by

$$Ef(\mathscr{C}_k) = E \sum_{F_k \in \mathscr{F}_k(\mathscr{C}_d)} f(F_k)/EN_k(\mathscr{C}_d) \qquad (3.2.10)$$

when   f   is a translation invariant non-negative measurable function.   □

<u>Proof</u>.   Let $z(x|\Phi) = z(F(x|\Phi))$ as in the proof of Proposition 3.2.1. From (3.2.4), (3.2.7), (3.2.9), and using the translation invariance of   f   and the equivariance of the centroid, it follows that

$$\lambda_k \, Ef(\mathscr{C}_k)$$

$$= \lambda_k \int_B Ef(\mathscr{C}_k)\,dy \quad \text{with} \quad |B|_d = 1$$

$$= E \sum_{x \in \Phi_k} f(F(x|\Phi))\,1(z(F(x|\Phi)) \in B)$$

$$= E \sum_{x_i \in \Phi} \sum_{F_k \in \mathscr{F}_k(C(x_i|\Phi))} f(F_k)\,1(z(F_k) \in B)/(d-k+1)$$

$$= \lambda_d \int E \sum_{F_k \in \mathscr{F}_k(\mathscr{C}_d+y)} f(F_k)\,1(z(F_k) \in B)\,dy/(d-k+1)$$

$$= \lambda \int E \sum_{F_k \in \mathscr{F}_k(\mathscr{C}_d)} f(F_k)\,1(z(F_k)+y \in B)\,dy/(d-k+1)$$

$$= \lambda \, E \sum_{F_k \in \mathscr{F}_k(\mathscr{C}_d)} f(F_k)/(d-k+1)$$

$$= \lambda_k \, E \sum_{F_k \in \mathscr{F}_k(\mathscr{C}_d)} f(F_k)/EN_k(\mathscr{C}_d)$$

which proves (3.2.10). □

Proposition 3.2.3. We have that

$$\lambda_0 = 2\lambda \ , \quad \lambda_1 = 3\lambda \ , \quad \lambda_2 = \lambda \quad \text{if} \quad d = 2 \ , \qquad (3.2.11)$$

while

$$\lambda_1 = 2\lambda_0 \ , \quad \lambda_2 = \lambda + \lambda_0 \ , \quad \lambda_3 = \lambda \quad \text{if} \quad d = 3 \ . \qquad (3.2.12)$$

□

Proof. We have already observed that $\lambda_d = \lambda$. Let $z(x|\Phi) = z(F(x|\Phi))$ be as in the proof of Proposition 3.2.1. Then using first Gram's relation (2.2.21) we find that

$$0 = \lambda \ E \sum_{k=0}^{d} (-1)^k \sum_{F_k \in \mathcal{F}_k(\mathcal{C}_d)} \phi(F_k, \mathcal{C}_d)$$

$$= \sum_{k=0}^{d} (-1)^k \lambda \int E \sum_{F_k \in \mathcal{F}_k(\mathcal{C}_d)} \phi(F_k, \mathcal{C}_d) 1(y \in B - z(F_k)) dy$$

with $|B|_d = 1$. Next by (3.2.4) with $z(x_i|\Phi) = x_i$,

$$0 =$$

$$\sum_{k=0}^{d} (-1)^k \ E \sum_{x_i \in \Phi} \sum_{F_k \in \mathcal{F}_k(C(x_i|\Phi) - x_i)} \phi(F_k, C(x_i|\Phi) - x_i) 1(x_i \in B - z(F_k))$$

Since $1(x_i \in B - z(F_k)) = 1(z(F_k + x_i) \in B)$ and $\mathcal{F}_k(C(x_i|\Phi) - x_i) = \mathcal{F}_k(C(x_i|\Phi)) - x_i$, we obtain that

$$0 = \sum_{k=0}^{d} (-1)^k \ E \sum_{x_i \in \Phi} \sum_{F_k \in \mathcal{F}_k(C(x_i|\Phi))} \phi(F_k, C(x_i|\Phi)) 1(z(F_k) \in B)$$

$$= \sum_{k=0}^{d} (-1)^k \ E \sum_{x \in \Phi_k} \{ \sum_{x_i \in x} \phi(F(x|\Phi), C(x_i|\Phi)) \} 1(z(F(x|\Phi)) \in B) \ .$$

Here $\{...\} = \sigma_d$, and hence by (3.2.5),

$$0 = \sum_{k=0}^{d} (-1)^k \lambda_k$$

or equivalently

$$\sum_{k=0}^{d-1} (-1)^k \lambda_k = (-1)^{d+1} \lambda \ . \tag{3.2.13}$$

Now, if $d = 2$, $N_0(\mathcal{C}_2) = N_1(\mathcal{C}_2)$, and so (3.2.9) implies that

$$3\lambda_0 = \lambda EN_0(\mathcal{C}_2) = \lambda EN_1(\mathcal{C}_2) = 2\lambda_1$$

which combined with (3.2.13) give (3.2.11). If $d = 3$, then Euler's relation (2.2.20) and (3.2.9) imply that

$$4\lambda_0 - 3\lambda_1 - 2\lambda_2 = \lambda E[N_0(\mathcal{C}_3) - N_1(\mathcal{C}_3) + N_2(\mathcal{C}_3)] = 2\lambda$$

which combined with (3.2.13) give (3.2.12). □

Remark 3.2.2. Note that in the proofs of Propositions 3.2.1-3.3.3 we used only the facts that the Voronoi tessellation is normal and its cells satisfy Euler's and Gram's relations. Therefore, these propositions remain true for arbitrary stationary normal tessellations of $\mathbb{R}^d$ having convex cells (see e.g. Møller, 1989).

Proposition 3.2.3 states that all intensities are determined if just one is known in the planar case or if two are known in the spatial case. For instance, consider a planar Voronoi tessellation observed within some bounded window B. Then an unbiased estimator of λ_0 is simply obtained by counting the number of vertices within B and dividing this by the area of B. Hence, using the linear relations (3.2.11) we obtain unbiased estimators of λ_1 and λ_2 too. □

Turning now to the Delaunay tessellation $\mathcal{J}(\Phi)$, intensities and Palm measures of the facets are defined analogously. Because of the duality we find that the intensity $\tilde{\lambda}_k$ of Delaunay k-facets is given by

$$\tilde{\lambda}_k = \lambda_{d-k}$$

and

$$\tilde{Q}_k(A) = E \sum_{x \in \Phi_{d-k}} 1(z(x|\Phi) \in B, \text{conv}\{x\} - z(x|\Phi) \in A) / (\tilde{\lambda}_k |B|_d) \quad (3.2.14)$$

becomes the Palm distribution of the typical Delaunay k-facet $\tilde{\mathcal{C}}_k$, $k = 0, \ldots, d$, cf. Exercise 3.3.3. The accompanying Campbell theorem states that

$$\tilde{\lambda}_k \int Ef(\tilde{\mathcal{C}}_k, y) dy = E \sum_{x \in \Phi_{d-k}} f(\text{conv}\{x\} - z(x|\Phi), z(x|\Phi)) \quad (3.2.15)$$

when f is a non-negative measurable function. It can be shown that the typical Delaunay cell and (d-1)-facet are related by

$$(d+1) Ef(\tilde{\mathcal{C}}_{d-1}) = E \sum_{F_{d-1} \in \mathcal{J}_{d-1}(\tilde{\mathcal{C}}_d)} f(F_{d-1}) \quad (3.2.16)$$

if f is translation invariant, cf. Exercise 3.2.4.

A natural choice of the centroid $z(x|\Phi)$ in this context could be the center $z(x)$ of the k-dimensional sphere which circumscribe the nuclei in $x \in \Phi_{d-k}$, cf. Lemma 2.1.1. Then

$$\widetilde{\mathscr{C}}_k = R_k \; conv\{U_{k0}, \ldots, U_{kk}\} \qquad (3.2.17)$$

where R_k is the radius of the circumsphere and $U_{k0}, \ldots, U_{kk} \in S^{d-1}$ are the directions to the vertices of $\widetilde{\mathscr{C}}_k$. For $k = 1$, $2R_1$ and $\overline{+}U_{10} = \pm U_{11}$ can be interpreted as the distance and direction between a typical pair of nuclei which are Voronoi neighbours. For $k = d$, R_d becomes the typical vertex-nucleus distance of the Voronoi tessellation. Furthermore, if $d = 2$, Proposition 2.1.7 establishes a one-to-one correspondence between $\{U_{20}, U_{21}, U_{22}\}$ and the directions of the three edges emanating from the typical Voronoi vertex, cf. (4.3.11) in Section 4.3.

Unbiased estimators of $\lambda_k \varrho_k$ and $\widetilde{\lambda}_k \widetilde{\varrho}_k$ suggest themselves using (3.2.6) and (3.2.14), respectively. However, in practice a tessellation may be observed within some bounded topological regular region only, so parts of the facets at the boundary are unobserved. This is a well-known problem in spatial statistics and stochastic geometry ('the problem of edge effects'). It might be managed by introducing certain correction terms in the estimator ('edge corrections') or by using a smaller window contained in the region ('minus sampling'), see e.g. Stoyan et al. (1987). Note that for the directions of edges emanating from those Voronoi vertices which are interior points of the observation window there are no problems with edge effects.

Finally, it should be noticed that the only specific probabilistic model, apart from very trivial and non-essential models, where explicit expressions of the Palm distributions of typical Voronoi and Delaunay facets are known, seem to be the homogeneous Poisson process considered in Chapter 4.

Exercise 3.2.1. Show step by step how (3.2.2)-(3.2.4) can be obtained.

Exercise 3.2.2. Prove that $\theta\mathscr{C}_k$ is distributed as $\mathscr{C}_k$ for $\theta \in \mathcal{O}(d)$ if Φ is isotropic and the centroid is equivariant under rotations.

Exercise 3.2.3. Verify in details why $\tilde{\lambda}_k = \lambda_{d-k}$ and show how (3.2.14)-(3.2.15) can be obtained.

Exercise 3.2.4. Prove (3.2.16) for $d = 2,3$.
Hint: $2\lambda_1 = (d+1)\lambda_0$ and each Delaunay $(d-1)$-facet is contained in precisely two Delaunay cells.

3.3. Mean value relations

Let the situation be as in Section 3.2. In this section we establish some mean value relations between certain characteristics of Voronoi facets.

The characteristics include those given by (2.2.16) and (2.2.17) with $X_k = \mathscr{C}_k$, $k = 0,\ldots,d$. Moreover, we consider the mean k-content of all Voronoi k-facets observed inside a window $B \subseteq \mathbb{R}^d$,

$$\Lambda_k(B) = E \sum_{x \in \Phi_k} |F(x|\Phi) \cap B|_k .$$

This is seen to be a translation invariant measure on the Borel sets $B \in \mathscr{B}$ and hence proportional to $|B|_d$. Clearly, $\Lambda_0(B) = \lambda_0 |B|_d$. For $d = 2$,

$$\Lambda_1(B) = L_A |B|_2 \tag{3.3.1}$$

where L_A is the expected boundary length of the planar Voronoi tessellation per unit area. Furthermore, if $d = 3$,

$$\Lambda_2(B) = S_V |B|_3 \qquad\qquad (3.3.2)$$

and

$$\Lambda_1(B) = L_V |B|_3 \qquad\qquad (3.3.3)$$

where S_V and L_V are the expected surface area and length of edges of the spatial Voronoi tessellation per unit volume, respectively. The quantities L_A, L_V, S_V are called the densities of edges in $\mathbb{R}^2$, edges in $\mathbb{R}^3$, and interfaces in $\mathbb{R}^3$, respectively. The relations (3.3.1)-(3.3.3) and the definition of $\Lambda_k(B)$ suggest immediately how these densities can be unbiased estimated without any problem with edge effects.

Proposition 3.3.1. If $d = 2$ then

$$EA(\mathscr{C}_2) = 1/\lambda \qquad\qquad (3.3.4)$$

$$EU(\mathscr{C}_2) = 2L_A/\lambda \qquad\qquad (3.3.5)$$

$$EN(\mathscr{C}_2) = 6 \qquad\qquad (3.3.6)$$

$$EL(\mathscr{C}_1) = L_A/3\lambda \; . \qquad\qquad (3.3.7)$$

If $d = 3$ then

$$EV(\mathscr{C}_3) = 1/\lambda \qquad\qquad (3.3.8)$$

$$ES(\mathscr{C}_3) = 2S_V/\lambda \qquad\qquad (3.3.9)$$

$$E\bar{b}(\mathcal{C}_3) = L_V/4\lambda \qquad\qquad (3.3.10)$$

$$E\bar{L}(\mathcal{C}_3) = 3L_V/\lambda \qquad\qquad (3.3.11)$$

$$EN_2(\mathcal{C}_3) = 2(1+\lambda_0/\lambda) \qquad\qquad (3.3.12)$$

$$EN_1(\mathcal{C}_3) = 6\lambda_0/\lambda \qquad\qquad (3.3.13)$$

$$EN_0(\mathcal{C}_3) = 4\lambda_0/\lambda \qquad\qquad (3.3.14)$$

$$EA(\mathcal{C}_2) = S_V/(\lambda+\lambda_0) \qquad\qquad (3.3.15)$$

$$EU(\mathcal{C}_2) = 3L_V/(\lambda+\lambda_0) \qquad\qquad (3.3.16)$$

$$EN(\mathcal{C}_2) = 6/(1+\lambda/\lambda_0) \qquad\qquad (3.3.17)$$

$$EL(\mathcal{C}_1) = L_V/2\lambda_0 \; . \qquad\qquad (3.3.18)$$

$$\square$$

<u>Proof</u>. For Borel sets $B \subseteq \mathbb{R}^d$ and $X_k \subseteq G_k$ with $G_k \in \mathscr{G}_k^{(d)}$ a k-flat, Fubini's theorem gives that

$$|B|_d|X_k|_k = \int \int 1(x^k \in X_k + y, y \in B) \, dx^k \, dy$$

$$\qquad\qquad (3.3.19)$$

$$= \int |(X_k+y) \cap B|_k \, dy \, , \qquad k = 0,\ldots,d \; .$$

Hence, using (3.2.7) and (3.3.19),

$$\Lambda_k(B) = \lambda_k \, E \int |(\mathcal{C}_k+y) \cap B|_k \, dy = \lambda_k |B|_d \, E|\mathcal{C}_k|_k$$

which combined with Proposition 3.2.3 and (3.3.1)-(3.3.3) imply that (3.3.4), (3.3.7), (3.3.8), (3.3.15), and (3.3.18) hold (note that $\Lambda_d(B) = |B|_d$).

Next, combine Propositions 3.2.1 and 3.2.3 to obtain (3.3.6) and (3.3.12)-(3.3.14). Then, if $d = 3$, (3.2.10) gives that

$$EN(\mathscr{C}_2) = E \sum_{F_2 \in \mathscr{F}_2(\mathscr{C}_3)} N(F_2)/EN_2(\mathscr{C}_3)$$

$$= 3\ EN_0(\mathscr{C}_3)/EN_2(\mathscr{C}_3)$$

as each vertex of $\mathscr{C}_3$ is contained in exactly three sides of $\mathscr{C}_3$, and so using (3.3.12) and (3.3.14) we obtain (3.3.17).

If $d = 2$ then (3.2.10) gives that

$$EU(\mathscr{C}_2) = E \sum_{F_1 \in \mathscr{F}_1(\mathscr{C}_2)} L(F_1) = EN(\mathscr{C}_2) EL(\mathscr{C}_1)$$

and so inserting (3.3.6) and (3.3.7) we get (3.3.5). Similarly, if $d = 3$, (3.2.10) implies that

$$ES(\mathscr{C}_3) = E \sum_{F_2 \in \mathscr{F}_2(\mathscr{C}_3)} A(F_2) = EN_2(\mathscr{C}_3) EA(\mathscr{C}_2)$$

and

$$E\bar{L}(\mathscr{C}_3) = E \sum_{F_1 \in \mathscr{F}_1(\mathscr{C}_3)} L(F_1) = EN_1(\mathscr{C}_3) EL(\mathscr{C}_1)$$

whereby (3.3.9) and (3.3.11) follow from (3.3.12)-(3.3.13), (3.3.15), and (3.3.18). Moreover, when $d = 3$, (3.2.10) gives that

$$EU(\mathscr{C}_2) = E \sum_{F_2 \in \mathscr{F}_2(\mathscr{C}_3)} U(F_2)/EN_2(\mathscr{C}_3) = 2 \ E\bar{L}(\mathscr{C}_3)/EN_2(\mathscr{C}_3)$$

as each edge of $\mathscr{C}_3$ is contained in exactly two sides of $\mathscr{C}_3$. Hence, inserting (3.3.11) and (3.3.12) proves (3.3.16).

Formulae (2.2.18) and (2.2.19) give that

$$E\bar{b}(\mathscr{C}_3) = \frac{1}{4\pi} \ E \sum_{F_1 \in \mathscr{F}_1(\mathscr{C}_3)} (\pi - \phi(F_1, \mathscr{C}_3)) L(F_1)$$

and so using (3.3.19) with $k = 1$, $|B|_3 = 1$, and $X_1 = F_1$,

$$E\bar{b}(\mathscr{C}_3) = \frac{1}{4} \ E\bar{L}(\mathscr{C}_3) - \frac{1}{4\pi} \int E \sum_{F_1 \in \mathscr{F}_1(\mathscr{C}_3)} \phi(F_1, \mathscr{C}_3) |(F_1+y) \cap B|_1 \ dy \ .$$

According to (3.2.4) the latter integral becomes

$$E \sum_{x_i \in \Phi} \sum_{F_1 \in \mathscr{F}_1(C(x_i|\Phi)+x_i)} \phi(F_1, C(x_i|\Phi)+x_i) |(F_1+x_i) \cap B|_1/\lambda$$

$$= E \sum_{x_i \in \Phi} \sum_{F_1 \in \mathscr{F}_1(C(x_i|\Phi))} \phi(F_1, (C(x_i|\Phi))) |F_1 \cap B|_1/\lambda$$

$$= E \sum_{x \in \Phi_1} \{ \sum_{x_i \in x} \phi(F(x|\Phi), C(x_i|\Phi)) \} |F(x|\Phi) \cap B|_1/\lambda$$

where $\{...\} = 2\pi$ as the Voronoi cells surround the edge $F(x|\Phi)$. Therefore,

$$E\bar{b}(\mathscr{C}_3) = \frac{1}{4} \ E\bar{L}(\mathscr{C}_3) - \frac{1}{2} \ E \sum_{x \in \Phi_1} |F(x|\Phi) \cap B|_1/\lambda$$

$$= \frac{1}{4} \ E\bar{L}(\mathscr{C}_3) - \frac{1}{2} \ L_V/\lambda$$

as $|B|_3 = 1$, and then inserting (3.3.11) we obtain (3.3.10). The proof is completed. □

Remark 3.3.1. The relationships stated in Proposition 3.3.1 show that all the mean value characteristics are parametrized by (λ, L_A) in the planar case and by $(\lambda, \lambda_0, L_V, S_V)$ in the spatial case. (Recall that the intensities $\lambda_0, \ldots, \lambda_d$ are also parametrized by λ if $d = 2$ and by (λ, λ_0) if $d = 3$, cf. Proposition 3.2.3.) The relationships are only due to the geometric structure of Voronoi tessellations, so it follows from the proof that Proposition 3.3.1 holds as well for arbitrary stationary normal tessellations having convex cells (even not convexity is essential for some of the relations to hold). Similar relationships exist without assuming normality, but then extra parameters are in general needed, see e.g. Stoyan et al. (1987, Ch. 10).

The Delaunay tessellation is not normal, but because of its duality to the Voronoi tessellation we obtain immediately from the proof of Proposition 3.3.1 that the following relationships hold (see Exercise 3.3.1): If $d = 2$ then

$$EA(\tilde{\mathscr{C}}_2) = 1/2\lambda \qquad (3.3.20)$$

so planar Voronoi cells are in average twice as large as planar Delaunay cells. Further, if $d = 2$,

$$EU(\tilde{\mathscr{C}}_2) = \tilde{L}_A/\lambda \qquad (3.3.21)$$

and

$$EL(\tilde{\mathscr{C}}_1) = \tilde{L}_A/3\lambda \qquad (3.3.22)$$

where $\tilde{L}_A$ is the density of Delaunay edges. Furthermore, if $d = 3$,

$$EV(\tilde{\mathscr{C}}_3) = 1/\lambda_0 \qquad (3.3.23)$$

$$ES(\tilde{\mathscr{C}}_3) = 2\tilde{S}_V/\lambda_0 \qquad (3.3.24)$$

$$EA(\tilde{\mathscr{C}}_2) = \tilde{S}_V/2\lambda_0 \qquad (3.3.25)$$

$$EL(\tilde{\mathscr{C}}_1) = \tilde{L}_V/(\lambda+\lambda_0) \qquad (3.3.26)$$

where $\tilde{L}_V$ and $\tilde{S}_V$ denote the densitites of Delaunay edges and interfaces, respectively. The expressions concerning $ES(\tilde{C}_3)$, $E\bar{b}(\tilde{\mathscr{C}}_3)$, $E\bar{L}(\tilde{\mathscr{C}}_3)$, and $EU(\tilde{\mathscr{C}}_2)$ when $d = 3$ turn out to be more complicated as certain multiplicities appear because the number of Delaunay cells containing a Delaunay edge is not constant.

□

The next proposition states an interesting relationship between the typical Voronoi cell and a 'point sampled' Voronoi cell.

<u>Proposition 3.3.2.</u> We have that

$$Ef(\mathscr{C}_d) = E[f(C_a(\Phi))/|C_a(\Phi)|_d]E|\mathscr{C}_d|_d \qquad (3.3.27)$$

for any translation invariant non-negative measurable function f. Here $C_a(\Phi)$ denotes the almost surely unique Voronoi cell which contains an arbitrary fixed point $a \in \mathbb{R}^d$. Especially,

$$E[A(C_2))^2] = E[A(C_a(\Phi))]/\lambda \quad \text{if} \quad d = 2 \qquad (3.3.28)$$

and

$$E[V(C_3))^2] = E[V(C_a(\Phi))]/\lambda \quad \text{if} \quad d = 3 . \qquad (3.3.29)$$

Proof. As the point a is contained in at least one Voronoi cell,

$$M := \sum_{x_i \in \Phi} 1(a \in C(x_i|\Phi)) \geq 1 .$$

But (3.2.4) gives that

$$EM = \lambda E \int 1(a \in \mathscr{C}_d + y) dy = \lambda E |\mathscr{C}_d|_d = 1 ,$$

cf. (3.3.4) and (3.3.8). Consequently M = 1 almost surely. This proves that $C_a(\Phi)$ exists with probability one. Note that the distribution of $C_a(\Phi)$ does not depend on $a \in \mathbb{R}^d$ as Φ is stationary.

Now, using (3.2.4) and the fact that f is translation invariant we see that

$$\lambda E[f(\mathscr{C}_d) |\mathscr{C}_d|_d]$$

$$= \lambda E \int f(\mathscr{C}_d) 1(-y \in \mathscr{C}_d) dy$$

$$= E \sum_{x_i \in \Phi} f(C(x_i|\Phi) - x_i) 1(-x_i \in C(x_i|\Phi) - x_i)$$

$$= E \sum_{x_i \in \Phi} f(C(x_i|\Phi)) 1(0 \in C(x_i|\Phi))$$

$$= E[f(C_0(\Phi))] .$$

Combining this with (3.3.4) and (3.3.8) we obtain (3.3.27). The special cases (3.3.28) and (3.3.29) follow immediately by setting $f(\cdot) = |\cdot|_d$. □

Remark 3.3.2. The cell $C_a(\Phi)$ is called a point-sampled cell as it is the cell containing the point a. Formula (3.3.27) relates its distribution P_d, say, to the Palm distribution Q_d of the typical Voronoi cell: As far as we are only concerned with translation invariant properties, P_d is proportional to Q_d weighted by the size of the typical cell. Therefore, P_d is also called the volume weighted distribution of an arbitrary selected Voronoi cell. Note that

$$E|\mathscr{C}_d|_d \leq E|C_a(\Phi)|_d \qquad (3.3.30)$$

with equality if and only if $|\mathscr{C}_d|_d$ is almost surely constant, see Exercise 3.3.2.

A fast sampling procedure for collecting planar cells is to use point sampling by means of a systematic grid of points. Then the cells containing these grid points are sampled; if a cell contains more than one grid point it is sampled repeatedly. The emperical distribution based on the point sampled cells approximates the distribution of $C_a(\Phi)$. Similarly, by weighting each point sampled cell by its size, we obtain an approximation of the distribution of the typical cell.

Exercise 3.3.1. Verify formulae (3.3.20)-(3.3.26).

Exercise 3.3.2. Verify formula (3.3.30) and show that equality holds exactly when $|\mathscr{C}_d|_d$ is constant with probability one.

Exercise 3.3.3 (Gabriel neighbours). The intensity of Gabriel (or full Voronoi) neighbours is defined as

$$\lambda_G = E \sum_{\{x_0,x_1\}\in\Phi_{d-1}} 1(\frac{x_0+x_1}{2} \in B\cap F(x_0,x_1|\Phi)) \qquad (3.3.31)$$

for an arbitrary Borel set $B \subset \mathbb{R}^d$ with $|B|_d = 1$, cf. (1.1.6). Show that

$$N_G = 2\lambda_G/\lambda \qquad (3.3.32)$$

can be given an interpretation as the mean number of Gabriel neighbours to the typical Voronoi cell.

This exercise is continued in Exercises 4.2.2. and 4.3.1.

3.4. Flat sections

In this section we consider the intersection of an arbitrary stationary Voronoi tessellation with a fixed n-flat $L_n \in \mathcal{G}_n^{(d)}$, $0 < n < d$. Examples can be seen in Section 4.5. Throughout this section the stationary point process Φ is assumed to have intensity $\lambda \in (0,\infty)$ and to be in general quadratic position with probability one.

The non-void sectional cells

$$C_n(x_i|\Phi) = C(x_i|\Phi)\cap L_n \qquad (3.4.1)$$

constitute a tessellation $\mathcal{V}_{L_n}(\Phi)$ of L_n as verified below. Recall that $\mathcal{V}_{L_n}(\Phi)$ is in general not a Voronoi tessellation, cf. Remark 2.1.6. Obviously, the distribution of $\mathcal{V}_{L_n}(\Phi)$ becomes invariant under translations in L_n. The aim of this section is to describe the intensities and typical facets of $\mathcal{V}_{L_n}(\Phi)$ and relate these to the intensities and typical facets of the Voronoi tessellation $\mathscr{V}(\Phi)$. The established relations become important for stereological applications (for a recent review on stereology, see Baddeley, 1991).

For ease of exposition, let us assume that the n-flat L_n contains the origin; since Φ is stationary this is no serious restriction. We may expect that non-void intersections of the form

$$F_n(x|\Phi) = \bigcap_{i=0}^{n-k} C_n(x_i|\Phi), \quad x = (x_0,\ldots,x_{n-k}) \in \Phi_{d-n+k},$$

should constitute the k-facets of the sectional tessellation. Indeed this is the case as verified below. Note that

$$F_n(x|\Phi) = F(x|\Phi) \cap L_n \qquad (3.4.2)$$

which is either the empty set or a polytope in L_n. In the following we assume that $\lambda_k < \infty$, $k = d-n,\ldots,d$, where λ_k is the intensity of k-facets of $\mathscr{V}(\Phi)$, cf. Section 3.2.

Proposition 3.4.1. With probability one we have that the aggregate $\mathscr{V}_{L_n}(\Phi)$ of non-empty sectional cells $C_n(x_i|\Phi)$, $x_i \in \Phi$, constitutes a normal tessellation of L_n. The k-facets of $\mathscr{V}_{L_n}(\Phi)$ are almost surely given by those $F_n(x|\Phi) \neq \emptyset$, $x \in \Phi_{d-n+k}$, and they coincide with the k-faces of the sectional cells, $k = 0,\ldots,n$. $\qquad\square$

Proof. For $x \in \Phi_{d-n+k}$ with $F_n(x|\Phi) \neq \emptyset$, let $\dim(F_n(x|\Phi))$ denote the dimension of the polytope $F_n(x|\Phi)$ if $k \geq 0$, and set $\dim(F_n(x|\Phi)) = -\infty$ if $k < 0$. Using (3.2.7) and (3.2.10) we have that

$$E \sum_{x \in \Phi_{d-n+k}} \chi(F_n(x|\Phi)) 1(\dim(F_n(x|\Phi)) \neq k)$$

$$= \lambda_{d-n+k} \int E[\chi((\mathscr{C}_{d-n+k}+y) \cap L_n)$$

$$1(\dim((\mathscr{C}_{d-n+k}+y) \cap L_n) \neq k)] dy$$

$$= \frac{\lambda_{d-n+k}}{EN_{d-n+k}(\mathscr{C}_d)} \int E \sum_{F_{d-n+k} \in \mathscr{F}_{d-n+k}(\mathscr{C}_d)} [\chi((F_{d-n+k}+y) \cap L_n)$$

$$1(\dim((F_{d-n+k}+y) \cap L_n) \neq k)] dy$$

$$= \frac{\lambda_{d-n+k}}{EN_{d-n+k}(\mathscr{C}_d)} \int_{L_n} E \sum_{F_{d-n+k} \in \mathscr{F}_{d-n+k}(\mathscr{C}_d)} \{ \int_{L_n^\perp} \chi((F_{d-n+k}+y^{d-n}) \cap L_n)$$

$$1(\dim((F_{d-n+k}+y^{d-n}) \cap L_n) \neq k) dy^{d-n}\} dy^n .$$

If k < 0 then

$$\chi((F_{d-n+k}+y^{d-n}) \cap L_n) = 0$$

except on a Lebesgue nullset with respect to dy^{d-n}. Therefore, if k < 0,

$$E \sum_{x \in \Phi_{d-n+k}} \chi(F_n(x|\Phi)) = 0$$

and so with probability one $F_n(x|\Phi) = \emptyset$ for all $x \in \Phi_{d-n+k}$.
On the other hand, if $k \geq 0$, then

$$1(\dim((F_{d-n+k}+y^{d-n}) \cap L_n) \neq k) = 0$$

except on a Lebesgue nullset with respect to dy^{d-n}, whereby $F_n(x|\Phi) \neq \emptyset$ implies that $\dim(F_n(x|\Phi)) = k$ almost surely for all $x \in \Phi_{d-n+k}$. Consequently, if $x \in \Phi_{d-n+k}$, $F_n(x|\Phi) \neq \emptyset$ implies that $k \in \{0,\ldots,n\}$ and $F_n(x|\Phi)$ is a k-polytope almost surely. The remaining part of the proposition is now straightforward verified.   □

As a matter of fact it can be shown that the intersection between L_n and an arbitrary stationary tessellation of bounded convex cells in $\mathbb{R}^d$ constitutes a tessellation of bounded convex cells of L_n, $0 < n < d$, where k-facets in L_n correspond to $(d-n+k)$-facets in $\mathbb{R}^d$. Indeed the above proof is easily modified to this general situation. The most important situation for stereological applications is of course the case $d = 3$ and $n = 2$. Then Proposition 3.4.1 simply states that the cells, edges, and vertices of the planar tessellation $\mathcal{V}_{L_2}(\Phi)$ correspond to those spatial cells, sides, and edges, respectively, which hit the plane L_2.

It is now straightforward to define intensities $\lambda_k^{(n)}$ and typical facets $\mathscr{C}_k^{(n)}$ with respect to $\mathcal{V}_{L_n}(\Phi)$, $k = 0,\ldots,n$. For that purpose it is convenient to define the centroid by

$$z_n(x|\Phi) = \Pi_{L_n}(z(x|\Phi)), \quad x \in \Phi_{d-n+k}^{(n)}, \qquad (3.4.3)$$

where $\Pi_{L_n}(z(x|\Phi))$ is the orthogonal projection of the centroid $z(x|\Phi) \in \mathbb{R}^d$ onto L_n, see Figure 3.4.1, and

$$\Phi_{d-n+k}^{(n)} = \{x \in \Phi_{d-n+k} | F_n(x|\Phi) \neq \emptyset\} \qquad (3.4.4)$$

is the collection of point configurations of nuclei which parametrize the k-facets of $\mathscr{V}_{L_n}(\Phi)$. Then $z^{(n)}(x|\Phi)$ beomes in-variant under translations in L_n. Further, the intensity

$$\lambda_k^{(n)} = E \sum_{x \in \Phi_{d-n+k}} 1(z_n(x|\Phi) \in B)/|B|_n \qquad (3.4.5)$$

is well-defined for an arbitrary Borel set $B \subset L_n$ with $0 < |B|_n < \infty$, and

$$\lambda_k^{(n)} \int_{L_n} Ef(\mathscr{C}_k^{(n)}, y^n) dy^n$$

$$(3.4.6)$$

$$= E \sum_{x \in \Phi_{d-n+k}^{(n)}} f(F_n(x|\Phi) - z_n(x|\Phi), z_n(x|\Phi))$$

for any non-negative measurable function f. As in Lemma 3.2.1 one obtains that $\lambda_k^{(n)}$ and $Ef(\mathscr{C}_k^{(n)})$ do not depend on the specific choice of the centroid $z_n(x|\Phi)$ provided f is in-variant under translations in L_n. However, the centroid (3.4.3) provides us easily to establish the following connection to the corresponding intensities and typical facets in $\mathbb{R}^d$.

Proposition 3.4.2. For $k = 0,\ldots,n$ we have that

$$\lambda_k^{(n)} = \lambda_{d-n+k} E|\Pi_{L_n^\perp}(\mathscr{C}_{d-n+k})|_{d-n} \qquad (3.4.7)$$

and

$$\lambda_k^{(n)} \; Ef(\mathscr{C}_k^{(n)})$$

$$(3.4.8)$$

$$= \lambda_{d-n+k} \; E \int_{\Pi_{L_n^\perp}(\mathscr{C}_{d-n+k})} f(\mathscr{C}_{d-n+k} \cap (y^{d-n}+L_n)) dy^{d-n}$$

whenever   f   is a nonnegative measurable function which is invariant under translations in $\mathbb{R}^d$.   □

   <u>Proof</u>.   Let $B \subset L_n$ with $|B|_n = 1$.   From (3.4.4) and (3.4.6),

$$\lambda_k^{(n)} \; Ef(\mathscr{C}_k^{(n)})$$

$$= E \sum_{x \in \Phi_{d-n+k}} \chi(F_n(x|\Phi)) f(F_n(x|\Phi)) 1(z_n(x|\Phi) \in B)$$

as χ and   f   are invariant under translations. Further, by (3.2.7) and (3.4.2)-(3.4.3),

$$= \lambda_{d-n+k} \int E[\chi((\mathscr{C}_{d-n+k}+y) \cap L_n) f((\mathscr{C}_{d-n+k}+y) \cap L_n)]$$

$$1(\Pi_{L_n}(y) \in B) dy$$

$$= \lambda_{d-n+k} \int_B dy^n \; E \int_{L_n^\perp} \chi(\mathscr{C}_{d-n+k} \cap (L_n-y^{d-n}))$$

$$f(\mathscr{C}_{d-n+k} \cap (L_n-y^{d-n})) dy^{d-n}$$

where we have used Fubini's theorem and the invariance of χ and   f   again. As $|B|_n = 1$ we obtain then (3.4.8), and hence the special case (3.4.7) by setting   f = 1.   □

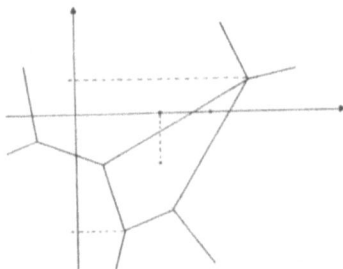

Figure 3.4.1. Part of a planar Voronoi tessellation and its intersection with a line L_1 (horizontal axis). The height of a Voronoi cell with respect to $L_1^\perp$ is indicated.

Remark 3.4.1. We can rewrite (3.4.8) as

$$Ef(\mathscr{C}_k^{(n)}) = \frac{\lambda_{d-n+k}}{\lambda_k^{(n)}} \; E \int_{\Pi_{L_n^\perp}(\mathscr{C}_{d-n+k})} |\Pi_{L_n^\perp}(\mathscr{C}_{d-n+k})|_{d-n}$$

$$f(\mathscr{C}_{d-n+k} \cap (y^{d-n}+L_n)) \; \frac{dy^{d-n}}{|\Pi_{L_n^\perp}(\mathscr{C}_{d-n+k})|_{d-n}} \; .$$

This means that as far as we are only concerned with translation invariant properties of $\mathscr{C}_k^{(n)}$, its distribution is determined by first drawing a facet F_{d-n+k} from the 'height' weighted distribution

$$\frac{\lambda_{d-n+k}}{\lambda_k^{(n)}} |\Pi_{L_n^\perp}(F_{d-n+k})|_{d-n} \; Q_{d-n+k}(dF_{d-n+k}) \qquad (3.4.9)$$

(see Figure 3.4.1) and then setting

$$\mathscr{C}_k^{(n)} = F_{d-n+k} \cap (y^{d-n} + L_n)$$

where y^{d-n} is uniformly distributed on $\Pi_{L_n^\perp}(F_{d-n+k})$.

This can be used for simulations. For instance, suppose that $d = 3$, $k = n = 2$, and we have an algorithm for simulating typical spatial Voronoi cells $\mathscr{C}_3$ (see Section 4.5). Then the typical cell $\mathscr{C}_2^{(2)}$ in the planar section can be simulated by first generating a sample $C_1, \ldots, C_m$ of realizations of $\mathscr{C}_3$, then for each C_i a uniform point y_i^1 on the interval $\Pi_{L_2^\perp}(C_i)$, and finally weighting each result $C_i^{(2)} = C_i \cap (y_i^1 + L_2)$ by

$$|\pi_{L_2^\perp}(C_i)|_1 / \sum_{j=1}^m |\pi_{L_2^\perp}(C_j)|_1 , \quad i = 1, \ldots, m .$$

Roughly speaking, the weighting by height just reflects that each Voronoi cell hits the plane L_2 with a change proportional to its height. Note that the height is simply determined by the vertices of the polyhedron $\mathscr{C}_3$.

Finally, note that Proposition 3.4.2 remains true for more general stationary processes of $(d-n+k)$-dimensional 'particles' intersected by an n-flat. □

We shall now use Proposition 3.4.2 to establish the connection between certain mean value characteristics of Voronoi facets and sectional facets. Again the results extend to more general processes as remarked above.

<u>Proposition 3.4.3</u>. Suppose that the typical Voronoi $(d-n+k)$-facet is isotropic. Then

$$\lambda_k^{(n)} \ E f(\mathscr{C}_k^{(n)}) = \lambda_{d-n+k} \ E \int_{\mathscr{G}_n^{(d)}} \chi(\mathscr{C}_{d-n+k} \cap G_n) f(\mathscr{C}_{d-n+k} \cap G_n) dG_n$$

(3.4.10)

for any nonnegative measurable function f which is invariant under motions in $\mathbb{R}^d$. Here dG_n denotes the motion invariant measure on the set $\mathscr{G}_n^{(d)}$ of n-flats in $\mathbb{R}^d$, cf. (2.2.6).

□

Proof. Let

$$g(L_n) = E \int_{L_n^\perp(\mathscr{C}_{d-n+k})} f(\mathscr{C}_{d-n+k} \cap (y^{d-n} + L_n)) dy^{d-n}$$

$$= \int_{L_n^\perp} E[\chi(\mathscr{C}_{d-n+k} \cap (y^{d-n} + L_n)) f(\mathscr{C}_{d-n+k} \cap (y^{d-n} + L_n))] dy^{d-n} \ .$$

Then for any orthogonal matrix $0 \in O(d)$ we find that

$$g(0 L_n) = \int_{0 L_n^\perp} E[\chi(\mathscr{C}_{d-n+k} \cap (y^{d-n} + 0 L_n))$$

$$f(\mathscr{C}_{d-n+k} \cap (y^{d-n} + 0 L_n))] dy^{d-n}$$

$$= \int_{L_n^\perp} E[\chi(0^* \mathscr{C}_{d-n+k} \cap (y^{d-n} + L_n))$$

$$f(0^* \mathscr{C}_{d-n+k} \cap (y^{d-n} + L_n)) dy^{d-n}$$

as χ and f are invariant under rotations and the Jacobian of the mapping $y^{d-n} \to 0 y^{d-n}$ is 1. Since $\mathscr{C}_{d-n+k}$ is distributed as $0^* \mathscr{C}_{d-n+k}$ we have that

$$g(L_n) = g(0L_n)$$

for all $0 \in 0(d)$. Hence, as the set of n-spaces

$$\mathscr{L}_n^{(d)} = \{0L_n | 0 \in 0(d)\}$$

it follows that $g(L_n)$ does not depend on $L_n \in \mathscr{L}_n^{(d)}$. Therefore, by (3.4.8),

$$\lambda_k^{(n)} \; Ef(\mathscr{C}_k^{(n)})$$

$$= \lambda_{d-n+k} \; g(L_n)$$

$$= \lambda_{d-n+k} \int_{\mathscr{L}_n^{(d)}} g(L_n) dL_n$$

$$= \lambda_{d-n+k} \int_{\mathscr{L}_n^{(d)}} \int_{L_n^\perp} E[\chi(\mathscr{C}_{d-n+k} \cap (y^{d-n}+L_n))$$

$$f(\mathscr{C}_{d-n+k} \cap (y^{d-n}+L_n))] dy^{d-n} \; dL_n$$

whereby Fubini's theorem and (2.2.6) imply (3.4.10). □

Proposition 3.4.4. The following relations hold for an isotropic Voronoi tessellation.

(i)   Case d = 2, n = 1:

$$\lambda_0^{(1)} = \lambda_1^{(1)} = 1/EL(\mathscr{C}_1^{(1)}) = 2L_A/\pi \qquad (3.4.11)$$

(ii)   Case d = 3, n = 1:

$$\lambda_0^{(1)} = \lambda_1^{(1)} = 1/EL(\mathscr{C}_1^{(1)}) = S_V/2 \qquad (3.4.12)$$

(iii) Case d = 3,   n = 2:

$$\lambda_0^{(2)} = \frac{1}{2} L_V \qquad (3.4.13)$$

$$\lambda_1^{(2)} = \frac{3}{4} L_V \qquad (3.4.14)$$

$$\lambda_2^{(2)} = \frac{1}{4} L_V \qquad (3.4.15)$$

$$EA(\mathscr{C}_2^{(2)}) = 4/L_V \qquad (3.4.16)$$

$$EU(\mathscr{C}_2^{(2)}) = 2\pi \ S_V/L_V \qquad (3.4.17)$$

$$EN(\mathscr{C}_2^{(2)}) = 6 \qquad (3.4.18)$$

$$EL(\mathscr{C}_1^{(2)}) = \frac{\pi}{3} \ S_V/L_V \ . \qquad (3.4.19)$$

$\square$

Proof. The proof is based on combining Crofton's sectional formulae (2.2.23)-(2.2.29) and the Propositions 3.2.3, 3.3.1, and 3.4.3.

(i)   Case d = 2, n = 1:

$$\lambda_1^{(1)} = \lambda_2 \ E \int x(\mathscr{C}_2 \cap G_1) dG_1 = \frac{\lambda_2}{\pi} \ EU(\mathscr{C}_2) = 2L_A/\pi$$

$$\lambda_1^{(1)} \ EL(\mathscr{C}_1^{(1)}) = \lambda_2 \ E \int L(\mathscr{C}_2 \cap G_1) dG_1 = \lambda_2 \ EA(\mathscr{C}_2) = 1$$

$$\lambda_0^{(1)} = \lambda_1 \ E \int x(\mathscr{C}_1 \cap G_1) dG_1 = \frac{\lambda_1}{\pi} \ EU(\mathscr{C}_1) = \frac{2\lambda_1}{\pi} \ EL(\mathscr{C}_1) = 2L_A/\pi$$

(ii) Case d = 3, n = 1: See Exercise 3.4.1.

(iii) Case d = 3, n = 2: See Exercise 3.4.1. $\square$

Remark 3.4.2. In Exercise 3.4.2 we rewrite the right hand sides of (3.4.11)-(3.4.19) as ratios between mean values of certain facet characteristics of the Voronoi tessellation. Note that (3.4.13)-(3.4.15) and (3.4.18) just reflect that the planar sectional tessellation is normal (compare with (3.2.11) and (3.3.6)). □

Proposition 3.4.5. Assuming isotropy then

$$\frac{2}{3} L_A E[L(\mathscr{C}_1^{(1)})^3] = \lambda E[A(\mathscr{C}_2)^2] \quad \text{if} \quad d = 2 , \quad (3.4.20)$$

$$\frac{2\pi}{3} S_V E[L(\mathscr{C}_1^{(1)})^3] = L_V E[A(\mathscr{C}_2^{(2)})^2] \quad \text{if} \quad d = 3 , \quad (3.4.21)$$

$$L_V E[L(\mathscr{C}_1^{(2)})^3] = (\lambda+\lambda_0) E[A(\mathscr{C}_2)^2] \quad \text{if} \quad d = 3 , \quad (3.4.22)$$

$$\frac{\pi}{6} S_V E[L(\mathscr{C}_1^{(1)})^4] = \lambda E[V(\mathscr{C}_3)^2] \quad \text{if} \quad d = 3 . \quad (3.4.23)$$

□

Proof. See Exercise 3.4.3. □

Let us consider again the case $d = 3$. Then by Proposition 3.4.4 it is obvious how to construct unbiased estimates of the densitites L_V and S_V based on a planar section containing a test set of systematic parallel and horizontal lines. The intercept lengths can also be used to obtain unbiased estimates of $E[L(\mathscr{C}_1^{(1)})^k]$, $k = 3,4$. Combining these estimates with the estimates of L_V and S_V, formulae (3.4.21) and (3.4.23) then suggest how one can obtain estimates of $E[A(\mathscr{C}_2^{(2)})^2]$ and $E[V(\mathscr{C}_3)^2]/E[V(\mathscr{C}_3)]$.

Finally, the next proposition extends Proposition 3.3.2 to the case of typical cells of $\mathscr{V}_{L_n}(\Phi)$.

Proposition 3.4.6. Let $L_n \in \mathscr{G}_n^{(d)}$ be an n-flat through a fixed point $a \in \mathbb{R}^d$, $0 < n < d$. Then

$$Ef(\mathscr{C}_n^{(n)}) = E[f(C_a(\Phi) \cap L_n)/|C_a(\Phi) \cap L_n|_n]E|\mathscr{C}_n^{(n)}|_n \quad (3.4.24)$$

for any non-negative measurable function f which is invariant under translations in L_n. □

Proof. See Exercise 3.4.4. □

Exercise 3.4.1. Prove the cases (ii) and (iii) of Proposition 3.4.4.

Exercise 3.4.2. Assuming that we have an isotropic Voronoi tessellation, establish the following relations by combining Propositions 3.3.1 and 3.4.4:

$$EL(\mathscr{C}_1^{(1)}) = \pi EA(\mathscr{C}_2)/EU(\mathscr{C}_2) \quad \text{if} \quad d = 2 , \quad (3.4.25)$$

$$EL(\mathscr{C}_1^{(1)}) = 4EV(\mathscr{C}_3)/ES(\mathscr{C}_3) \quad \text{if} \quad d = 3 , \quad (3.4.26)$$

$$EA(\mathscr{C}_2^{(2)}) = EV(\mathscr{C}_3)/E\bar{b}(\mathscr{C}_3) \quad \text{if} \quad d = 3 , \quad (3.4.27)$$

$$EU(\mathscr{C}_2^{(2)}) = \frac{\pi}{4} ES(\mathscr{C}_3)/E\bar{b}(\mathscr{C}_3) \quad \text{if} \quad d = 3 , \quad (3.4.28)$$

$$EN(\mathscr{C}_2^{(2)}) = \frac{1}{2} E\bar{L}(\mathscr{C}_3)/E\bar{b}(\mathscr{C}_3) \quad \text{if} \quad d = 3 , \quad (3.4.29)$$

$$EL(\mathscr{C}_1^{(2)}) = \pi EA(\mathscr{C}_2)/EU(\mathscr{C}_2) \quad \text{if} \quad d = 3 . \quad (3.4.30)$$

Moreover, using Proposition 3.3.1 and Proposition 3.4.5, show that

$$E[L(\mathscr{C}_1^{(1)})^3] = 3E[A(\mathscr{C}_2)^2]/EU(\mathscr{C}_2) \quad \text{if} \quad d = 2 , \quad (3.4.31)$$

$$E[L(\mathscr{C}_1^{(2)})^3] = 3E[A(\mathscr{C}_2)^2]/EU(\mathscr{C}_2) \quad \text{if} \quad d = 3 \ , \quad (3.4.32)$$

$$E[L(\mathscr{C}_1^{(1)})^4] = \frac{12}{\pi} E[V(\mathscr{C}_3)^2]/ES(\mathscr{C}_3) \quad \text{if} \quad d = 3 \ . \quad (3.4.33)$$

Explain the correspondence between these formulae and those in Exercise 2.2.5.

Remark 3.4.3. In stereology, when one uses random probes like the IUR-lines and IUR-planes introduced in Exercise 2.2.4, the approach is said to be design-based, while the approach used in the foregoing of this section is said to be model-based. Formulae (3.4.25)-(3.4.33) remain true in the case where the Voronoi tessellation is anisotropic, but the sectioning line and plane are IUR.

Exercise 3.4.3. Prove Proposition 3.4.5 by combining Propositions 2.2.2, 3.4.3 and 3.4.4.

Exercise 3.4.4. Extend the proof of Proposition 3.3.2 to the situation considered in Proposition 3.4.6.

4. Poisson-Voronoi tessellations

The homogeneous Poisson point process on $\mathbb{R}^d$ seems to be the only non-trivial stochastic model for which a reasonable collection of theoretical results for the associated Voronoi tessellation is derivable. In Section 4.1 we introduce Poisson point processes and study some of their properties. Especially, we state Slivnyak's theorem which makes Poisson-Voronoi tessellation tractable for mathematical and statistical analysis as shown in Sections 4.2-4.4. The results presented in Section 4.2 on mean value characteristics are mostly due to Meijering (1953) and Gilbert (1962), while the results in Sections 4.3 and 4.4 on the distributions of typical Poisson-Voronoi and Poisson-Delaunay statistics are mainly due to Collins (1968), Miles (1970, 1974, 1984) and unpublished research by Brakke (1987a, 1987b). We shall present and prove all results in Sections 4.2-4.4 using the approach of Palm measure theory as introduced in Chapter 3 (see also Møller, 1989). Finally, in Section 4.5 we present an algorithm due to Quine and Watson (1984) for simulating the homogeneous Poisson process and other related models. The algorithm is tailormade for simulating the typical Voronoi cell as well as the Voronoi tessellation within a ball avoiding any problem with edge effects. Some empirical results are discussed.

4.1. The homogeneous Poisson process

A Poisson point process is roughly speaking a model for a spatial point process Φ with no interaction between its points. Below we present some equivalent definitions and derive a few useful properties when Φ is supposed to be stationary.

Recall that we consider only spatial point processes $\Phi \subset \mathbb{R}^d$ which has no multiple points, cf. Section 3.1. There is a lack of interaction between the points in Φ if $N_{B_1}(\Phi), \ldots, N_{B_n}(\Phi)$ are independent for any disjoint bounded

Borel sets $B_1, \ldots, B_n \subset \mathbb{R}^d$, $n \geq 2$, where $N_B(\Phi) = \#(\Phi \cap B)$. This property together with stationarity can be shown to imply that

$$P(\Phi \cap K = \emptyset) = \exp(-\lambda |K|_d) \qquad (4.1.1)$$

for any compact set $K \subset \mathbb{R}^d$, where $\lambda \geq 0$ is a constant (the case $\lambda = 0$ is trivial, because then $\Phi = \emptyset$ almost surely). Since the void-probabilities uniquely determine the distribution of Φ it follows that the following properties must hold for any bounded Borel set $B \subset \mathbb{R}^d$ (see Exercise 4.1.2):

(i) $N_B(\Phi)$ is Poisson distributed with mean $\lambda |B|_d$.

(ii) Given that $N_B(\Phi) = n$, the n points $\Phi \cap B = \{x_1, \ldots, x_n\}$ form a binomial process on B, i.e. $x_1, \ldots, x_n$ are independent and uniformly distributed in B.

On the other hand, it is easily seen that these conditions specify a well-defined distribution of a stationary point process on $\mathbb{R}^d$ with intensity λ. By (i)-(ii) we have also isotropy for this process. If $\lambda > 0$ then the restriction of the process to any topological regular set $S \subseteq \mathbb{R}^d$ is called a homogeneous Poisson point process on S. This definition can be extended to the case of inhomogeneous Poisson point processes by replacing the mean measure $\lambda |B|_d$ in (i) by an arbitrary measure $\Lambda(B)$ which is finite for all bounded Borel sets B.

Proposition 4.1.1 (Slivnyak's theorem). Let Φ be a homogeneous Poisson point process on $\mathbb{R}^d$ with intensity $\lambda > 0$. Then, for $n = 1, 2, \ldots,$

$$E \sum_{\{x_1,\ldots,x_n\}\subseteq\Phi} f(\{x_1,\ldots,x_n\},\Phi)$$

(4.1.2)

$$= \frac{\lambda^n}{n!} \int\ldots\int Ef(\{y_1,\ldots,y_n\},\Phi\cup\{y_1,\ldots,y_n\})dy_1\ldots dy_n$$

for any non-negative measurable function f, where the sum is over all subsets $\{x_1,\ldots,x_n\}\subseteq\Phi$ of n distinct nulcei.

□

Proof. Let us first consider the case $n = 1$. By standard measure theoretical methods it suffices to prove that

$$E \sum_{x_i\in\Phi\cap B} 1[N_K(\Phi)=0] = \lambda \int_B P(N_K(\Phi\cup\{y\}=0)dy$$

whenever $K \subset \mathbb{R}^d$ is compact and $B \subset \mathbb{R}^d$ is a bounded Borel set, cf. the comments concerning the σ-field $\mathcal{F}$ in Section 3.1. Since $N_K(\Phi)$ and $N_{B\setminus K}(\Phi)$ are independent, property (i) gives that

$$E \sum_{x_i\in\Phi\cap B} 1[N_K(\Phi)=0]$$

$$= E(N_{B\setminus K}(\Phi)1[N_K(\Phi)=0])$$

$$= E(N_{B\setminus K}(\Phi))P(N_K(\Phi)=0)$$

$$= \lambda|B\setminus K|_d P(N_K(\Phi)=0)$$

$$= \lambda \int_{B\setminus K} P(N_K(\Phi)=0)dy$$

$$= \lambda \int_B P(N_K(\Phi\cup\{y\})=0)dy ,$$

whereby the case $n = 1$ is proved. The proof for $n > 1$ follows now by induction (Exercise 4.1.3). □

Remark 4.1.1. The Palm distribution of a homogeneous Poisson process Φ is obtained by combining (3.1.7) with (4.1.2). Using the stationarity of Φ this reduces to

$$Q(F) = P(\Phi \cup \{0\} \in F) , \qquad (4.1.3)$$

see Exercise 4.1.4. Note that Q is isotropic but not stationary. In fact, (4.1.3) characterizes the homogeneous Poisson process, cf. e.g. Jagers (1973). □

Remark 4.1.2. It follows from the proof of Proposition 4.1.1 that (4.1.2) holds as well for an inhomogeneous Poisson point process with mean measure Λ provided $\lambda^n \, dy_1 \ldots dy_n$ is replaced by the product measure $\Lambda(dy_1) \ldots \Lambda(dy_n)$ (if Λ is not a diffuse measure, i.e. absolutely continuous with respect to Lebesgue measure, then the sum in (4.1.2) is over all configurations of n not necessarily distinct nuclei). □

Proposition 4.1.2. A homogeneous Poisson point process on $\mathbb{R}^d$ satisfies the conditions (a), (b), (c) (see Sections 1.1 and 2.1) with probability one. □

Proof. Using Proposition 4.1.1 we see that

$$E \sum_{\{x_0, \ldots, x_k\} \subseteq \Phi} 1[x_0, \ldots, x_k \text{ lie on a } (k-1)\text{-flat}]$$

$$= \frac{\lambda^{k+1}}{(k+1)!} \int \ldots \int 1[x_0, \ldots, x_k \text{ lie on a } (k-1)\text{-flat}] dx_0 \ldots dx_k$$

$$= 0, \quad k = 2, \ldots, d ,$$

whereby (a) is seen to hold almost surely. Similarly,

$$E \sum_{\{x_1,\ldots,x_{d+2}\}\subseteq\Phi} 1[x_1,\ldots,x_{d+2} \text{ lie on the boundary of a}$$

$$\text{sphere}]$$

$$= \frac{\lambda^{d+2}}{(d+2)!} \int\cdots\int 1[x_1,\ldots,x_{d+2} \text{ lie on the boundary of}$$

$$\text{a sphere}]dx_1\ldots dx_{d+2}$$

$$= 0$$

which proves that (b) holds almost surely. Finally, Proposition 3.1.1 gives that (c) holds almost surely.   □

Exercise 4.1.1. Let Φ_n be a binomial process of N_n points on a bounded Borel set $B_n \subset \mathbb{R}^d$, $n = 1,2,\ldots$. Assume that B_n contains the ball $B(0,r_n) = \{y\in\mathbb{R}^d | \|y\|<r_n\}$ where $r_n \to \infty$ as $n \to \infty$. Finally, assume that

$$N_n/|B_n|_d \to \lambda \quad \text{as} \quad n \to \infty$$

where $\lambda \in (0,\infty)$. Show that

$$\lim_{n\to\infty} P(\Phi_n\cap K=\emptyset) = \exp(-\lambda|K|_d)$$

for any compact set $K \subset \mathbb{R}^d$ (compare with (4.1.1)).

Exercise 4.1.2. Verify that if (4.1.1) holds for all compact sets $K \subset \mathbb{R}^d$, then (i)-(ii) must be satisfied for all bounded Borel sets $B \subset \mathbb{R}^d$.

Hint to (ii): The void-probabilities for the binomial process with n points in B are given by

$$\omega_K = \left[\frac{|B|_d - |B \cap K|_d}{|B|_d} \right]^n$$

provided $0 < |B|_d < \infty$.

Exercise 4.1.3. Assuming that (4.1.2) holds for $n = m-1$ with $m \geq 2$, show that (4.1.2) is satisfied when $n = m$.

Exercise 4.1.4. Verify that (4.1.3) becomes the Palm distribution of a homogeneous Poisson point process Φ.

Hint: $P(\Phi \cup \{0\}) = P((\Phi \cup \{y\}) - y \in F)$.

4.2. Mean value characteristics of Poisson-Voronoi facets

Throughout this section and Sections 4.3-4.4, Φ denotes a homogeneous Poisson process on $\mathbb{R}^d$ with intensity λ. Hence the Voronoi tessellation $\mathscr{V}(\Phi)$ and the Delaunay tessellation $\mathscr{T}(\Phi)$ become both stationary and isotropic. In this section we determine the measures

$$\Lambda_{d-k}(B) = E \sum_{x \in \Phi_{d-k}} |F(x|\Phi) \cap B|_{d-k}, \quad k = 1, \ldots, d ,$$

which in turn determine the densities L_A ($d=2$) and λ_0, L_V, S_V ($d=3$), cf. Section 3.3. Inserting the values of these densities in the relations given in Propositions 3.2.3, 3.3.1, and 3.4.4 we obtain all intensities and first order moments of the characteristics of $\mathscr{V}(\Phi)$ as well as the sectional Voronoi tessellations $\mathscr{V}_{L_n}(\Phi)$, $0 < n < d$, when $d = 2$ and $d = 3$. Moreover, in this section, we show how the variance of the size distributions for typical cells of $\mathscr{V}(\Phi)$ and $\mathscr{V}_{L_n}(\Phi)$ can be determined.

We turn first to the densities. Using Proposition 4.1.1 we obtain that

$$\Lambda_{d-k}(B) = \frac{\lambda^{k+1}}{(k+1)!} \int \ldots \int E|F(x|\Phi\cup x)\cap B|_{d-k} \, dx_0 \ldots dx_k$$

where $x = \{x_0, \ldots, x_k\}$, and so by Fubini's theorem and Lemma 2.1.1 we have that

$$\Lambda_{d-k}(B)$$

$$= \frac{\lambda^{k+1}}{(k+1)!} \int \ldots \int \int_{L_{d-k}} P(y^{d-k} + z \in F(x|\Phi\cup x)\cap B) \, dy^{d-k} \, dx_0 \ldots dx_k$$

$$(4.2.1)$$

with $L_{d-k} = L(x)^{\perp}$ and $z = z(x)$ (in the special case $k = d$, dy^0 denotes the Dirac measure concentrated at the origin $L_0 = \{0\}$). Now, making the variable shift (2.2.34) from x to $(z, R, u_0, \ldots, u_k, L_k)$ we find that

$$y^{d-k} + z \in F(x|\Phi\cup x)\cap B$$

$$\Updownarrow$$

$$z \in B - y^{d-k}, \quad B(z+y^{d-k}, r)\cap\Phi = \emptyset$$

where $B(a,r)$ is the ball in $\mathbb{R}^d$ with center a and radius

$$r = \{R^2 + \|y^{d-k}\|^2\}^{1/2} ,$$

see also Figure 4.2.1. Therefore,

$$\int_{L_{d-k}} P(y^{d-k} + z \in F(x|\Phi\cup x)\cap B) \, dy^{d-k}$$

$$= \int_{L_{d-k}} 1(z \in B - y^{d-k}) \, \upsilon(r) \, dy^{d-k}$$

where $\mathfrak{v}(r)$ is the void-probability

$$\mathfrak{v}(r) \equiv P(\Phi \cap B(0,r)) = \exp(-\lambda \omega_d r^d) \qquad (4.2.2)$$

and where

$$\omega_d = \pi^{d/2}/\Gamma\left(\tfrac{d}{2} + 1\right)$$

is the volume of the unitball in $\mathbb{R}^d$. Inserting this into (4.2.1), using the Blaschke-Petkantschin formula (2.2.35), and integrating over $z \in B-y^{d-k}$ and L_k, where $\int dL_k = 1$, we have that

$$\Lambda_{d-k}(B) = \frac{\lambda^{k/d}}{(k+1)!} \overset{'}{c}_k^{(d)} I_k J_k |B|_d \qquad (4.2.3)$$

with

$$I_k = \lambda^{k+1-k/d} \int\limits_0^\infty \int\limits_{R^{d-k}} R^{dk-1} \mathfrak{v}(r) dy^{d-k} dR$$

and

$$J_k = \int\limits_{S^{k-1}} \cdots \int\limits_{S^{k-1}} v^{d-k+1} v_k(du_0) \cdots v_k(du_k) . \qquad (4.2.4)$$

For $k < d$ the integral I_k is evaluated by shifting the variables (y^{d-k}, R) to $(u, r, t) \in S^{d-k-1} \times (0, \infty) \times (0, 1)$ with

$$y^{d-k} = r\sqrt{1-t}\, u , \qquad R = r\sqrt{t} .$$

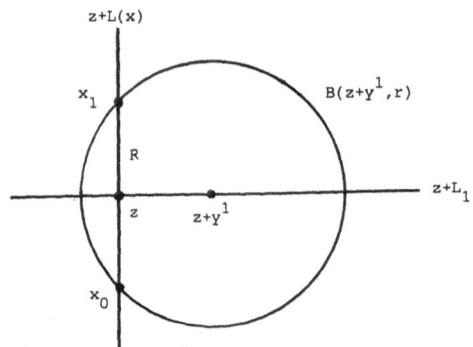

<u>Figure 4.2.1.</u>   The case $d = 2$ and $k = 1$.

Then

$$dy^{d-k} \; dR = \frac{1}{2} \; r^{d-k} \; t^{-1/2} (1-t)^{\frac{d-k}{2} - 1} \; v_{d-k}(du) \, dr \, dt$$

so

$$\lambda^{\frac{k}{d}-k-1} \; I_k = \frac{\sigma_{d-k}}{2} \int_0^\infty r^{dk+d-k-1} \; v(r) \, dr \int_0^1 t^{\frac{dk}{2} - 1} (1-t)^{\frac{d-k}{2} - 1} \; dt$$

where

$$\sigma_d = 2\pi^{d/2} / \Gamma\left(\frac{d}{2}\right)$$

is the surface area of the unit sphere in $\mathbb{R}^d$.   For $\alpha, \beta, \gamma > 0$
it is well-known that

$$\int_0^\infty r^{\alpha-1} \; \exp(-\gamma r^\beta) \, dr = \frac{\Gamma(\alpha/\beta)}{\beta \gamma^{\alpha/\beta}} \qquad (4.2.5)$$

and

$$\int_0^\infty t^{\alpha-1}(1-t)^{\beta-1}\ dt = \Gamma(\alpha)\Gamma(\beta)/\Gamma(\alpha+\beta)\ . \qquad (4.2.6)$$

Therefore, for all $k = 1,\ldots,d-1$,

$$I_k = \pi^{(d-k)/2}\frac{\Gamma(k+1-k/d)}{d\omega_d^{k+1-k/d}}\frac{\Gamma(dk/2)}{\Gamma(\frac{dk+d-k}{2})}\ . \qquad (4.2.7)$$

It is straightforward verified that (4.2.7) remains true when $k = d$.

Hence it remains only to evaluate the integral J_k given by (4.2.4). For this we need the following lemma.

Lemma 4.2.1. For integers $k \geq l \geq 1$, let Δ be the l-content of the simplex with vertices at $l+1$ independent and uniformly distributed points on S^{k-1}. Then the moments $c(\alpha,k,l) = E\Delta^\alpha$, $\alpha = 1,2,\ldots$, are given by

$$c(\alpha,k,l) = (\frac{1}{l!})^\alpha\frac{\Gamma(\frac{1}{2}(l+1)(k+\alpha)-1)}{\Gamma(\frac{1}{2}[(l+1)k+l\alpha]-1)}$$

$$\qquad (4.2.8)$$

$$\left\{\frac{\Gamma(\frac{k}{2})}{\Gamma(\frac{1}{2}[k+\alpha])}\right\}^{l+1}\prod_{j=1}^{l}\frac{\Gamma(\frac{1}{2}[k-l+\alpha+j])}{\Gamma(\frac{1}{2}[k-l+j])}$$

□

Proof. See Miles (1971, Theorem 2); see also Exercise 4.2.1. □

Letting $k = 1$ we have that $v = k!\Delta$, and so (4.2.4) and Lemma 4.2.1 imply that

$$J_k = \sigma_k^{k+1}(k!)^{d-k+1} c(d-k+1,k,k)$$

$$= 2^{k+1} \pi^{k(k+1)/2} \frac{\Gamma(\frac{1}{2}(k+1)(d+1)-k)}{\Gamma(\frac{kd}{2})\Gamma(\frac{d+1}{2})^{k+1}}$$

$$\prod_{j=1}^{k} \frac{\Gamma(\frac{1}{2}[d-k+1+j])}{\Gamma(j/2)} \quad.$$

Combining this with (4.2.3) and (4.2.7) we find that

$$L_A = 2\lambda^{1/2} \quad \text{if} \quad d = 2 , \tag{4.2.9}$$

$$\lambda_0 = \frac{24}{35} \pi^2 \lambda \simeq 6.768 \lambda \quad \text{if} \quad d = 3 , \tag{4.2.10}$$

$$L_V = \frac{16}{15}(\tfrac{3}{4})^{1/3} \pi^{5/3} \Gamma(\tfrac{4}{3})\lambda^{2/3} \simeq 5.832 \lambda^{2/3} \quad \text{if} \quad d = 3 , \tag{4.2.11}$$

$$S_V = 4(\pi/6)^{1/3} \Gamma(5/3)\lambda^{1/3} \simeq 2.910 \lambda^{1/3} \quad \text{if} \quad d = 3 , \tag{4.2.12}$$

for the cases $k = 1,3,2,1$, respectively. Inserting these values into the formulae in Propositions 3.2.3, 3.3.1, and 3.4.4 we obtain the numerical values of intensities and first order moments for facets as exhibited in Tables 4.2.1 and 4.2.2. Observe that the moments of $\mathscr{V}(\Phi)$ when $d = 2$ disagree with the corresponding moments of $\mathscr{V}_2(\Phi)$ when $d = 3$. This is in accordance with Remark 2.1.6: A planar section through a spatial Voronoi tessellation is not a planar Voronoi tessellation.

Having determined 'all' first order moments, let us next turn to higher order moments. As an illuminating example we consider the second order moment $E[|\mathscr{C}_n^{(n)}|_n^2]$ when $0 < n \leq d$

and $d = 2,3$, where $\mathscr{C}_d^{(d)} = \mathscr{C}_d$ denotes the typical Voronoi cell and $\mathscr{C}_n^{(n)}$ is the typical cell of $\mathscr{V}_{L_n}(\Phi)$ if $n < d$. As a biproduct, using Proposition 3.4.5, we obtain also the moments $E[L(\mathscr{C}_1^{(1)})^3]$ when $d = 2,3$ and $E[A(\mathscr{C}_2)^2]$ and $E[L(\mathscr{C}_1^{(1)})^4]$ when $d = 3$.

Now, to find $E[|\mathscr{C}_n^{(n)}|_n^2]$ is equivalent to determine the mean n-content of $C_0(\Phi) \cap L_n$ as $E[|\mathscr{C}_n^{(n)}|_n]$ is known, cf. Propositions 3.3.2 and 3.4.6. For this we proceed as in Gilbert (1962): Using Fubini's theorem,

$$E|C_0(\Phi) \cap L_n|_n$$

$$= \int_{L_n} P(y^n \epsilon C_0(\Phi)) dy^n$$

$$= \sigma_n \int_0^\infty P(s) s^{n-1} ds \qquad (4.2.13)$$

where

$$P(s) = P(su_0 \epsilon C_0(\Phi))$$

for an arbitrary unit vector u_0 as Φ is isotropic. Hence by Slivnyak's theorem (4.1.2),

$$P(s) = E \sum_{x_i \epsilon \Phi} 1(\{0, su_0\} \subset C(x_i | \Phi))$$

$$= \lambda \int P(\{0, su_0\} \subset C(y | \Phi \cup \{y\})) dy$$

$$= \lambda \int_0^\infty \int P(\{0, su_0\} \subset C(tu | \Phi \cup \{tu\}) t^{d-1} v_d(du) dt . \qquad (4.2.14)$$

In order to evaluate this integral the following lemma is convenient.

Lemma 4.2.2. For integers $d > k \geq 1$ and $u \in S^{d-1}$ we have that

$$v_d(du) = r^{k-1}(1-r^2)^{(d-k-2)/2} dr \, v_k(d\theta) v_{d-k}(d\phi) \quad (4.2.15)$$

where $u = (r\theta, \sqrt{1-r^2} \phi)$ with range $0 < r < 1$, $\theta \in S^{k-1}$, $\phi \in S^{d-k-1}$. □

Proof. Let $f = f(r,\theta,\phi) \geq 0$ be any measurable function and use (4.2.5) to obtain that

$$\frac{1}{2} \Gamma(d/2) \int f \, v_d(du)$$

$$= \int_0^\infty R^{d-1} e^{-R^2} dR \int f \, v_d(du)$$

$$= \int_{\mathbb{R}^d} e^{-\|y\|^2} f \, dy$$

$$= \int_{\mathbb{R}^k} dy_1 \int_{\mathbb{R}^{d-k}} dy_2 \, e^{-\|y_1\|^2 - \|y_2\|^2} f$$

$$= \int v_k(d\theta) \int v_{d-k}(d\phi) \int_0^\infty dr_1 \int_0^\infty dr_2 \, e^{-r_1^2 - r_2^2} r_1^{k-1} r_2^{d-k-1}$$

$$f(r_1(r_1^2+r_2^2)^{-1/2}, u_1, u_2) .$$

Substituting $r_1 = r\sqrt{t}$ and $r_2 = \sqrt{1-r^2} \sqrt{t}$ by (r,t) we get further that

$$= \frac{1}{2} \int_0^\infty t^{(d-2)/2} e^{-t} dt \iiint r^{k-1} (1-r^2)^{(d-k-2)/2} f(r,\theta,\phi)$$

$$dr \; v_k(d\theta) v_{d-k}(d\phi)$$

as the Jacobian is $(2\sqrt{1-r^2})^{-1}$. But according to (4.2.5) we have also that

$$\Gamma(d/2) = \int_0^\infty t^{(d-2)/2} e^{-t} dt$$

whereby (4.2.15) is seen to hold. □

Now, letting $k = 1$ in (4.2.15) we see from (4.2.14) that

$$P(s) = \lambda \int_0^\infty \int_0^1 \int_{S^{d-2}} \sum_{\theta \in \{\pm 1\}} t^{d-1} (1-r^2)^{(d-3)/2}$$

$$P(\{0, su_0\} \subset \subset ((tr\theta, t\sqrt{1-r^2} \phi) \mid \Phi U\{(tr\theta, t\sqrt{1-r^2} \phi)\})$$

$$v_{d-1}(d\phi) dr \; dt$$

and so taking $u_0 = (1,0,\ldots,0)^*$ and using that Φ is isotropic,

$$P(s) = \lambda \; \sigma_{d-1} \int_0^\infty \int_{-1}^1 t^{d-1} (1-r^2)^{(d-3)/2} P(\{0, su_0\} \subset \subset (y \mid \Phi U\{y\})) dr \; dt$$

$$(4.2.16)$$

with

$$y = tru_0 + t\sqrt{1-r^2} u_1, \quad u_1 = (0,\ldots,0,1)^*,$$

see Figure 4.2.2. Note that $\{0, su_0\} \subset C(y | \Phi \cup \{y\})$ if and only if no nuclei are contained in the union of the balls $B(0, \|y\|)$ and $B(su_0, \|y - su_0\|)$, where $\|y\| = t$ and

$$\|y - su_0\|^2 = \|y\|^2 + \|su_0\|^2 - 2sy \cdot u_0$$

$$= t^2 + s^2 - 2str .$$

Consequently, the void-probability

$$P(\{0, su_0\} \subset C(y | \Phi \cup \{y\}) = \exp(-\lambda \ast (r, s, t)) \qquad (4.2.17)$$

where

$$\ast(r, s, t) = |B(0, \|y\|) \cup B(su_0, \|y - su_0\|)|_d .$$

Since

$$\ast(r, s, t) = s^d \ast(r, 1, t/s)$$

we obtain from (4.2.13) and (4.2.16) that

$$E|C_0(\Phi) \cap L_n|_n$$

$$= \lambda \, \sigma_n \, \sigma_{d-1} \int_0^\infty \int_0^\infty \int_{-1}^1 s^{n-1} t^{d-1} (1-r^2)^{(d-3)/2} \exp(-\lambda s^d \ast(r, 1, t/s))$$

$$drdsdt$$

and so by Fubini's theorem, substituting t/s by q,

$$= \lambda \, \sigma_n \, \sigma_{d-1} \int_0^\infty \int_{-1}^1 \int_0^\infty q^{d-1} (1-r^2)^{(d-3)/2} s^{d+n-1}$$

$$\exp(-\lambda s^d \ast(r, 1, q)) dsdrdq$$

and hence by using (4.2.5),

$$E|C_0(\Phi) \cap L_n|_n$$

$$= \frac{\sigma_n \, \sigma_{d-1}}{d\lambda^{n/d}} \, \Gamma(\frac{d+n}{d}) \int_0^\infty \int_{-1}^1 \frac{q^{d-1}(1-r^2)^{(d-3)/2}}{\mathfrak{w}(r,1,q)^{(d+n)/d}} \, dr dq \quad (4.2.18)$$

where $\mathfrak{w}(r,1,q)$ is the content of the union of two balls with unit distance between the centers and radii q and $\sqrt{q^2+1-2qr}$, respectively. Finally, (4.2.18) can be evaluated by numerical integration, whereby the moments $E[|\mathfrak{C}_n^{(n)}|_n^2]$, $0 < n \le d$, are determined.

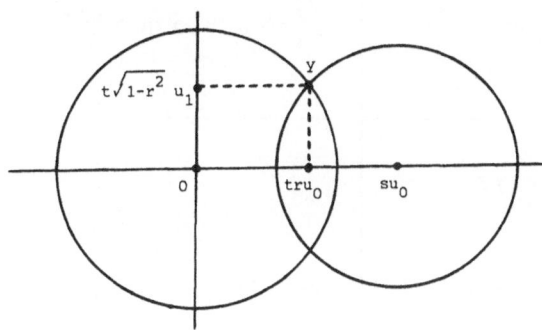

Figure 4.2.2. The case $d = 2$.

If one wants to determine the second order moments of other statistics like e.g. the total k-content of all k-faces included in $\mathfrak{C}_d$, $0 \le k < d$, the bookkeeping becomes even more laborious and void-probabilities appear which are at least as complicated as that in (4.2.17). Also covariances between the statistics can be included and integrals written down, but in general the details are messy. This is demonstrated in two incredible but unpublished papers by Brakke (1987 a,b).

Gilbert's and Brakke's results are summarized in Tables 4.2.3-4.2.7 below by means of the dimensionless quantities CV and ρ. Here

$$CV(X) = \sqrt{\text{Var}(X)}/E(X)$$

is the coefficient of variation of a random variable X, while

$$\rho(X,Y) = \text{Cov}(X,Y)/\sqrt{\text{Var}(X)\,\text{Var}(Y)}$$

is the correlation coefficient of two random variables X and Y.

$\lambda_0 = 2\lambda$ $\lambda_1 = 3\lambda$ $\lambda_2 = \lambda$	$\lambda_0^{(1)} \simeq 1.273\,\lambda^{1/2}$ $\lambda_1^{(1)} \simeq 1.273\,\lambda^{1/2}$
$EA(\mathscr{C}_2) = \lambda^{-1}$ $EU(\mathscr{C}_2) = 4\lambda^{-1/2}$ $EN(\mathscr{C}_2) = 6$ $EL(\mathscr{C}_1) \simeq 0.667\,\lambda^{-1/2}$	$EL(\mathscr{C}_1^{(1)}) \simeq 0.785\,\lambda^{-1/2}$

Table 4.2.1. Intensities and first order moments for a planar Poisson-Voronoi tessellation and for its intersection with a line.

$\lambda_0 \simeq 6.768 \lambda$	$\lambda_0^{(2)} \simeq 2.916 \lambda^{2/3}$	$\lambda_0^{(1)} \simeq 1.455 \lambda^{1/3}$
$\lambda_1 \simeq 13.535 \lambda$,	$\lambda_1^{(2)} \simeq 4.374 \lambda^{2/3}$	$\lambda_1^{(1)} \simeq 1.455 \lambda^{1/3}$
$\lambda_2 \simeq 7.768 \lambda$	$\lambda_2^{(2)} \simeq 1.458 \lambda^{2/3}$	
$\lambda_3 = \lambda$		
$EV(\mathscr{C}_3) = \lambda^{-1}$		
$ES(\mathscr{C}_3) \simeq 5.821 \lambda^{-2/3}$		
$E\bar{b}(\mathscr{C}_3) \simeq 1.458 \lambda^{-1/3}$		
$E\bar{L}(\mathscr{C}_3) \simeq 17.496 \lambda^{-1/3}$		
$EN_2(\mathscr{C}_3) \simeq 15.54$		
$EN_1(\mathscr{C}_3) \simeq 40.61$		
$EN_0(\mathscr{C}_3) \simeq 27.07$		
$EA(\mathscr{C}_2) \simeq 0.375 \lambda^{-2/3}$	$EA(\mathscr{C}_2^{(2)}) \simeq 0.686 \lambda^{-2/3}$	
$EU(\mathscr{C}_2) \simeq 2.252 \lambda^{-1/3}$	$EU(\mathscr{C}_2^{(2)}) \simeq 3.136 \lambda^{-1/3}$	
$EN(\mathscr{C}_2) \simeq 5.228$	$EN(\mathscr{C}_2^{(2)}) = 6$	
$EL(\mathscr{C}_1) \simeq 0.4309 \lambda^{-1/3}$	$EL(\mathscr{C}_1^{(2)}) \simeq 0.526 \lambda^{-1/3}$	$EL(\mathscr{C}_1^{(1)}) \simeq 0.687 \lambda^{-1/3}$

Table 4.2.2. Intensities and first order moments for a spatial Poisson-Voronoi tessellation and for its intersections with a plane and a line.

k	d = 2	d = 3
2	$0.806 \ \lambda^{-1}$	$0.632 \ \lambda^{-2/3}$
3	$0.960 \ \lambda^{-3/2}$	$0.668 \ \lambda^{-1}$
4		$0.774 \ \lambda^{-4/3}$

Table 4.2.3. Higher order moments $E[L(C_1^{(1)})^k]$ for the typical intercept length of planar and spatial Poisson-Voronoi tessella-tions.

	$A(\mathscr{C}_2)$	$U(\mathscr{C}_2)$	$N(\mathscr{C}_2)$	$L(\mathscr{C}_1)$
CV	0.529	0.243	0.222	0.646
$U(\mathscr{C}_2)$	0.953			
$N(\mathscr{C}_2)$	0.568	0.501		

Table 4.2.4. CV's and ρ's for the typical planar Poisson-Voronoi cell and edge.

	$A(\mathscr{C}_2)$	$U(\mathscr{C}_2)$	$N(\mathscr{C}_2)$	$L(\mathscr{C}_1)$
CV	1.007	0.538	0.302	0.753
$U(\mathscr{C}_2)$	0.928			
$N(\mathscr{C}_2)$	0.750	0.746		

Table 4.2.5. CV's and ρ's for the typical spatial Poisson-Voronoi side and edge.

	$A(\mathscr{C}_2^{(2)})$	$U(\mathscr{C}_2^{(2)})$	$N(\mathscr{C}_2^{(2)})$	$L(\mathscr{C}_1^{(2)})$
CV	0.695	0.388	0.282	0.691
$U(\mathscr{C}_2^{(2)})$	0.941			
$N(\mathscr{C}_1^{(2)})$	0.753	0.727		

Table 4.2.6. CV's and ρ's for the typical planar sectional Poisson-Voronoi cell and edge.

	$V(\mathscr{C}_3)$	$S(\mathscr{C}_3)$	$\bar{L}(\mathscr{C}_3)$	$N_0(\mathscr{C}_3)$
CV	0.423	0.254	0.211	0.246
$S(\mathscr{C}_3)$	0.982			
$\bar{L}(\mathscr{C}_3)$	0.938	0.924		
$N_0(\mathscr{C}_3)$	0.736	0.712	0.875	

Table 4.2.7. CV's and ρ's for the typical spatial Poisson-Voronoi cell.

Exercise 4.2.1. Verify Lemma 4.2.1 in the special case where $l = 1$.

Hint: Since the surface measure is invariant under rotations,

$$c(\alpha,k,1) = \int_{S^{k-1}} \int_{S^{k-1}} \|u_0-u_1\|^\alpha \, \nu_k(du_0)\nu_k(du_1)$$

$$= \sigma_k \int_{S^{k-1}} \|v-u\| \nu_k(du)$$

with $v = (1,0,\ldots,0)^* \in \mathbb{R}^k$. The case $k = l = 1$ is now straightforward. Lemma 4.2.2 can be used for $k > 1$.

Exercise 4.2.2 (Gabriel neighbours - continued). Let the situation be as in Exercise 3.3.3. Show that the mean number of Gabriel neighbours to the typical Poisson-Voronoi cell becomes

$$N_G = 2^d . \tag{4.2.18}$$

4.3. On the distribution of the typical Poisson-Delaunay cell and related statistics

Below we specify the distribution of the typical Poisson-Delaunay cell. Then this is used to determine the distributions of various related statistics as introduced at the end of Section 3.2.

Proposition 4.3.1. Consider the typical Poisson-Delaunay cell $\tilde{\mathcal{C}}_d = R_d \ \mathrm{conv}\{U_{d0}, \ldots, U_{dd}\}$, cf. (3.2.17). Then the typical vertex-nucleus distance R_d follows a generalized Gamma distribution with density

$$\frac{d(\lambda\omega_d)^d}{(d-1)!} \ r^{d^2-1} \ \exp(-\lambda\omega_d r^d\} \ , \qquad r > 0 \ . \qquad (4.3.1)$$

Moreover, the directions $(U_{d0}, \ldots, U_{dd})$ to the nuclei of the typical Voronoi vertex have joint density

$$\frac{\Delta(u_0, \ldots, u_d)}{\sigma_d^{d+1} \ c(1,d,d)} \ \nu_d(du_0) \cdots \nu_d(du_d) \ , \qquad u_i \in S^{d-1} \ , \qquad i = 0, \ldots, d \ ,$$

$$(4.3.2)$$

with $c(1,d,d)$ and $\Delta(u_0, \ldots, u_d)$ as defined in Lemma 4.2.1. Finally, R_d and $(U_{d0}, \ldots, U_{dd})$ are stochastic independent.

□

Proof. Since the centroid $z(x|\Phi)$ is chosen to be the center $z(x)$ of the open ball $B(x)$ containing $x = \{x_0, \ldots, x_d\}$ in its boundary, and $x \in \Phi_0$ if and only if $B(x) \cap \Phi = \emptyset$, (3.2.14) and Slivnyak's theorem (4.1.2) imply that

$$\tilde{Q}_d(A) = \frac{\lambda^{d+1}}{\lambda_0 |B|_d} \int \ldots \int 1(z(x) \in B, \text{ conv}\{x\} - z(x) \in A)$$

$$\Delta R(x)^{d^2-1} \exp(-\lambda \omega_d R(x)) dx_0 \ldots dx_d$$

where $R(x)$ is the radius of $B(x)$. Hence by Blaschke-Pet-kantschin's formula (2.2.35) we obtain that $\tilde{Q}_d(A)$ is proportional to

$$\int\int \ldots \int 1(R \text{conv}\{u_0, \ldots, u_d\} \in A) \Delta R^{d^2-1} \exp(-\lambda \omega_d R) v_d(du_0) \ldots v_d(du_d) dR$$

which combined with (4.2.5) and (4.2.8) prove the assertions.

□

Since Proposition 4.3.1 completely specifies the distribution of the typical Poisson-Delaunay cell it should at least in principle be possible to derive the distributions of various related statistics. For example,

$$|\tilde{\mathscr{C}}_d|_d = R_d^d \Delta (U_{0d}, \ldots, U_{dd})$$

so it is straightforward to derive all moments of the size distribution of the typical Poisson-Delaunay cell by using (4.2.5), (4.2.8), and Proposition 4.3.1. The results become

$$E[|\tilde{\mathscr{C}}_2|_2^\alpha] = \frac{\Gamma(\frac{3\alpha+5}{2})\Gamma(\frac{\alpha}{2}+1)}{(2\lambda)^\alpha \ 3\Gamma(\frac{\alpha+3}{2})^2 \ \pi^{\alpha-\frac{1}{2}}} \qquad \text{if} \quad d = 2 , \qquad (4.3.3)$$

and

$$E[|\tilde{\mathscr{C}}_3|_3^\alpha] = \frac{35\pi^{1/2}(\alpha+1)!(\alpha+2)!(2\alpha+4)!}{(16\pi\lambda)^\alpha \ 256 \ \Gamma(\frac{\alpha}{2}+2)^3 \ \Gamma(\frac{9+3\alpha}{2})} \qquad \text{if} \quad d = 3 , \qquad (4.3.4)$$

for $\alpha = 1, 2, \ldots$. These moments can be used to derive the probability density functions of $|\tilde{\mathscr{C}}_d|_d$, see Rathie (1992). In the planar case one obtains the density

$$\frac{8}{9} \pi \lambda^2 v K_{1/6}^2 \left[\frac{2\pi\lambda v}{3\sqrt{3}} \right] \;, \qquad v > 0 \;, \qquad\qquad (4.3.5)$$

where $K_{1/6}(\cdot)$ denotes the modified Bessel function of order 1/6. The result becomes much more complicated when $d = 3$.

In the remaining part of this section we restrict attention to the planar case $d = 2$.

Let us first consider the directions $U_{2i} = (\cos \theta_i, \sin \theta_i)$, $i = 0, 1, 2$, to the vertices $R_2 U_{12}$, $i = 0, 1, 2$, of the typical Poisson-Delaunay cell $\tilde{\mathscr{C}}_2$. Let β_i denote the angle of the triangle $\tilde{\mathscr{C}}_2$ at the vertex $R_2 U_{12}$. Then with θ_0 fixed there is a 2 to 1 correspondence between (θ_1, θ_2) and (β_1, β_2) with Jacobian

$$\left| \frac{\partial (\beta_1, \beta_2)}{\partial (\theta_1, \theta_2)} \right|_* = \frac{1}{4} \; .$$

This can be seen by using (4.3.11) and (4.3.12) below. Further,

$$\Delta = 2 \sin \beta_0 \sin \beta_1 \sin \beta_2$$

so it follows from Proposition 4.3.1 that $R_2, \theta_0, (\beta_1, \beta_2)$ are mutually independent with the following marginal densities:

$$2\pi^2 \lambda^2 r_2^3 \exp(-\pi\lambda r_2^2) \;, \qquad r_2 > 0 \;, \qquad\qquad (4.3.6)$$

and

$$\frac{8}{3\pi} \sin \beta_0 \sin \beta_1 \sin \beta_2 \;, \qquad 0 < \beta_i < \pi \;, \qquad i = 0, 1, 2 \;, \qquad (4.3.7)$$

with $\beta_0 = \pi - \beta_1 - \beta_2$, and θ_0 is uniformly distributed between 0 and 2π. We can interpret $R_2, \theta_0, (\beta_1, \beta_2)$ as the size, orientation, and shape, respectively, of the cell $\tilde{\mathscr{C}}_2$. From (4.3.7) we obtain that the probability of an obtuse-angled Delaunay triangle is 1/2. Further, the density of the typical angle $\beta = \beta_i$ between two Delaunay edges emanating from a nucleus becomes

$$\frac{4}{3\pi} \{(\pi - \beta)\cos \beta + \sin \beta\}\sin \beta , \quad 0 < \beta < \pi , \quad (4.3.8)$$

from which we obtain that

$$E(\beta) = \frac{\pi}{3} , \quad \mathrm{Var}(\beta) = \frac{\pi^2}{9} - \frac{5}{6} , \quad P(\beta < \frac{\pi}{2}) = \frac{5}{6} . \quad (4.3.9)$$

The density (4.3.8) is shown in Figure 4.3.1.

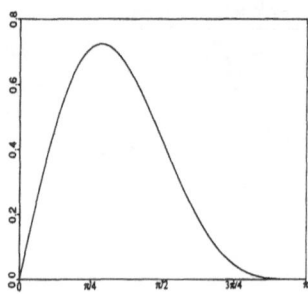

<u>Figure 4.3.1</u>. The density of the typical Delaunay triangle angle β.

Next, let us consider the typical Delaunay edge $\tilde{\mathscr{C}}_1$. This is distributed as a randomly picked edge of $\tilde{\mathscr{C}}_2$ where each edge has probability 1/3 for being selected, cf. (3.2.16). Therefore, the direction U_{10} of $\tilde{\mathscr{C}}_1$ is isotropic and stochastic independent of the length $2R_1 = |\tilde{\mathscr{C}}_1|_1$, cf. (3.2.17). Here R_1 is stochastic equivalent to $R_2 \sin \beta$. Using (4.3.6) and (4.3.8) it can then be shown that R_1 has density

$$\frac{4}{3} t(1-\Phi(t) + \sqrt{2\pi\lambda}\ t\phi(t)), \quad r_1 > 0 , \qquad (4.3.10)$$

with $t = r_1\sqrt{2\pi\lambda}$ and where $\phi(\cdot)$ and $\Phi(\cdot)$ denote the density and distribution functions of the standard normal distribution. Recall that R_2 and $2R_1$ can be interpreted as the typical vertex-nucleus distance and the typical nucleus-nucleus distance, respectively, cf. Section 3.2. Their density functions are shown in Figures 4.3.2 and 4.3.3. We have that

$$E(R_2) = \frac{3}{4\lambda^{1/2}} = 0.75\ \lambda^{-1/2} ,$$

$$\text{Var}(R_2) = (\frac{2}{\pi} - \frac{9}{16})/\lambda \simeq 0.0741\ \lambda^{-1} ,$$

$$E(2R_1) = \frac{32}{9\pi\lambda^{1/2}} \simeq 1.1318\ \lambda^{-1/2} ,$$

$$\text{Var}(2R_1) = (\frac{5}{\pi} - (\frac{32}{9\pi})^2)/\lambda \simeq 0.3106\ \lambda^{-1} .$$

Finally, in order to describe the distribution of the directions of the three edges meeting at a typical Voronoi vertex, let us order the angles $\theta_0, \theta_1, \theta_2$ such that $0 < \theta_1 < \theta_0 < \theta_2 < 2\pi$, see Figure 4.3.4. Using Proposition 2.1.7 we can always find these angles from the three Voronoi edges. Further, let γ be the angle of the Voronoi edge between θ_1 and θ_2 and let α_1, α_2 be the angles between that edge and the

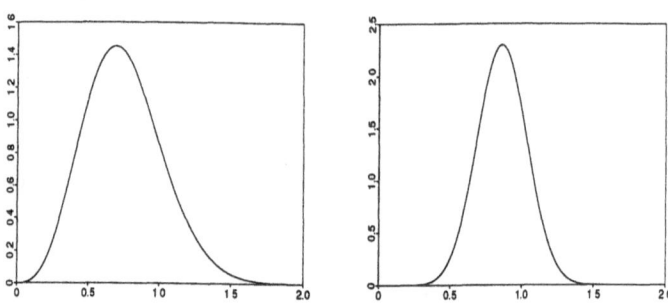

Figure 4.3.2. Density of the typical vertex-nucleus distance R_d when $\lambda = 1$. Left: $d = 2$. Right: $d = 3$.

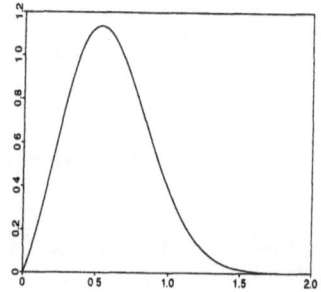

Figure 4.3.3. Density of the typical nucleus-nucleus distance $2R_1$ when $\lambda = 1$ and $d = 2$.

remaining two Voronoi edges as shown in Figure 4.3.4. Then by Proposition 2.1.7,

$$\alpha_1 = \pi - \frac{\theta_2 - \theta_0}{2} \ , \quad \alpha_2 = \pi - \frac{\theta_0 - \theta_1}{2} \ , \quad \gamma = \frac{\theta_1 + \theta_2}{2} \ , \qquad (4.3.11)$$

which establishes a one-to-one correspondence between $(\theta_0, \theta_1, \theta_2)$ and $(\alpha_1, \alpha_2, \gamma)$ with

$$d\theta_0 d\theta_1 d\theta_2 = 4 \ d\gamma \ d\alpha_1 \ d\alpha_2 .$$

It is easily shown that

$$\alpha_i + \beta_i = \pi , \qquad i = 0,1,2, \tag{4.3.12}$$

so

$$\Delta = 2 \sin \alpha_0 \sin \alpha_1 \sin \alpha_2$$

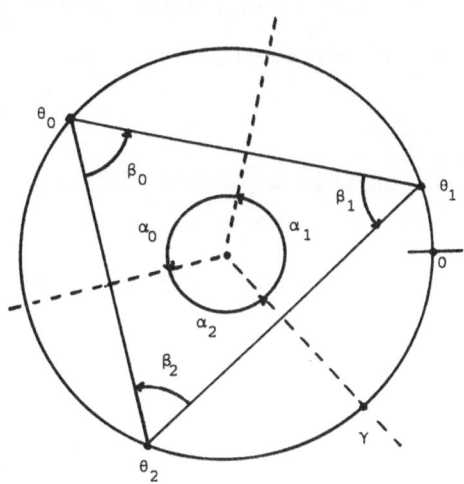

Figure 4.3.4. Angles $0 < \theta_1 < \theta_0 < \theta_2 < 2\pi$ to nuclei, angles β_i between Delaunay edges, angles α_i between Voronoi edges, and the angle γ between θ_1 and θ_2. All these angles are determined uniquenly by the three Voronoi edges.

with

$$\alpha_0 = 2\pi - \alpha_1 - \alpha_2 .$$

Consequently, (4.3.2) implies that

$$\frac{4}{\pi^2} \sin \alpha_0 \sin \alpha_1 \sin \alpha_2 \qquad (4.3.13)$$

becomes the joint density of $(\alpha_1, \alpha_2, \gamma)$. Because of the con-
straint $0 < \theta_1 < \theta_2 < 2\pi$ the range of $(\alpha_1, \alpha_2, \gamma)$ is

$$0 < \alpha_i < \pi , \quad i = 0,1,2, , \quad \alpha_0 < \gamma < \alpha_1 + \alpha_2 .$$

Hence given (α_1, α_2) the conditional distribution of γ is
uniform and symmetric about 0, and the joint density of
(α_1, α_2) becomes

$$\frac{8}{\pi^2} (\alpha_1 + \alpha_2) \sin \alpha_0 \sin \alpha_1 \sin \alpha_2 . \qquad (4.3.14)$$

Equivalently, because of the relation (4.3.12),

$$\frac{8}{\pi^2} \beta_0 \sin \beta_0 \sin \beta_1 \sin \beta_2 \qquad (4.3.15)$$

is the joint density of (β_1, β_2). This is just the density
(4.3.7) weighted by β_0.

Exercise 4.3.1 (Gabriel neighbours - continued). Let the
situation be as in Exercises 3.3.3 and 4.2.2. State an appropriate definition of the length $\mathcal{L}$ of the typical Gabriel edge
when Φ is a stationary point process in $\mathbb{R}^d$. Show that $\mathcal{L}$
follows a generalized Gamma density given by

$$\lambda d 2^{-d} \omega_d l^{d-1} \exp(-\lambda 2^{-d} \omega_d l^d) , \quad l > 0 , \qquad (4.3.16)$$

if Φ is a homogeneous Poisson process on $\mathbb{R}^d$ with intensity λ.

4.4. On the distribution of the typical Poisson-Voronoi cell and related statistics

Let Φ be a homogeneous Poisson process with intensity λ on $\mathbb{R}^d$. In this section we describe first the distribution of the typical Poisson-Voronoi cell $\mathscr{C}_d$ and secondly the distributions of various related statistics. It turns out that these distributions are mathematically less tractable than that of the typical Poisson-Delaunay cell and its related statistics as considered in Section 4.3.

Proposition 4.4.1. Let the centroids of the Voronoi cells be given by their nuclei, i.e., $z(x_i | \Phi) = x_i$ for all $x_i \in \Phi$. Then $\mathscr{C}_d$ is distributed as $C(0 | \Phi \cup \{0\})$. □

Proof. See Remark 4.1.1. □

Let us see how far we can proceed, using this characterization of the typical Voronoi cell.

The cell $C(0 | \Phi \cup \{0\})$ is completely determined by the neighbouring nuclei to 0 with respect to $\mathscr{V}(\Phi \cup \{0\})$. For $\Phi = \{x_i\}$ define

$$\mathscr{N}(x_i | \Phi) = \{x_j \in \Phi | j \neq i, F(x_i, x_j | \Phi) \neq \emptyset\} , \qquad (4.4.1)$$

the "$\mathscr{V}(\Phi)$-neighbourhood to x_i", which consists of a least $d+1$ nuclei. From the discussion in Section 1.1 on the Delaunay graph it follows that for $x_0 \in \Phi$ and $x = \{x_1, \ldots, x_n\} \subseteq \Phi$

with $x_0 \notin x$ and $n \geq d+1$,

$$N(x_0 | \Phi) = x$$

$$\Updownarrow$$

$$\bigcup_{\{x_1, \ldots, x_d\} \subseteq x} B(x_0, x_1, \ldots, x_d) \cap \Phi = \emptyset \qquad (4.4.2)$$

where $B(x_0, \ldots, x_d)$ is the open ball containing the points $x_0, \ldots, x_d$ in its boundary.

Now, using Proposition 4.4.1 and Slivnyak's theorem (4.1.2), we have that

$$P(\mathscr{C}_d \in A, \ N_{d-1}(\mathscr{C}_d) = n)$$

$$= E \sum_{x = \{x_1, \ldots, x_n\} \subseteq \Phi} 1(C(0 | \Phi \cup \{0\}) \in A, \ N(0 | \Phi \cup \{0\}) = x)$$

$$= \frac{\lambda^n}{n!} \int \ldots \int P(C(0 | \Phi \cup x \cup \{0\}) \in A,$$

$$N(0 | \Phi \cup x \cup \{0\}) = x) dx_1 \ldots dx_n$$

with $x = \{x_1, \ldots, x_n\}$ and $n \geq d+1$. According to (4.4.2), $N(0 | \Phi \cup x \cup \{0\}) = x$ if and only if $N(0 | x \cup \{0\}) = x$ and Φ has no points in

$$\mathscr{G}(x) = \bigcup_{\{x_1, \ldots, x_d\} \subseteq x} B(0, x_1, \ldots, x_d) , \qquad (4.4.3)$$

in which case $C(0 | \Phi \cup x \cup \{0\}) = C(0 | x \cup \{0\})$. (Note that $\mathscr{G}(x)$ is the union of balls circumscribing the Delaunay cells of $\mathscr{T}(x \cup \{0\})$ with a vertex at 0, see Figure 4.4.1). Consequently,

$$P(\mathscr{C}_d \in A, \ N_{d-1}(\mathscr{C}_d) = n)$$

(4.4.4)

$$= \frac{\lambda^n}{n!} \int \dots \int 1(C(0|x \cup \{0\}) \in A, \ \mathscr{N}(0|x \cup \{0\}) = x)$$

$$\exp(-\lambda |\mathscr{G}(x)|_d) dx_1 \dots dx_n \ .$$

It seems impossible to reduce this further. Even the simplest case, the probability of $\mathscr{C}_2$ being a triangle when $d = 2$, involves laborious calculations which do not provide much insight. Instead a Monte Carlo approach must be preferred; see Section 4.5.

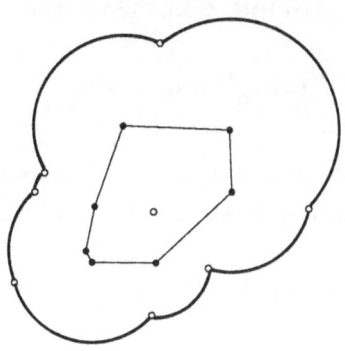

<u>Figure 4.4.1</u>. The planar case $d = 2$. The union of discs constitutes the void-area (4.4.3).

<u>Remark 4.4.1</u>. Let us show one special application of (4.4): Let $\mathscr{G}$ be the d-dimensional content of the union of all balls centered at the vertices of $\mathscr{C}_d \equiv C(0|\Phi \cup \{0\})$ and containing the nucleus 0 in their boundaries, see Figure 4.4.1. Then

$$\mathcal{G} = |\mathcal{G}(x)|_d \quad \text{if} \quad \mathcal{N}(0|\Phi U\{0\}) = x ,$$

so (4.4.4) implies that $P(\mathcal{G} \in B | N_{d-1}(\mathcal{C}_d) = n)$ is proportional to

$$\int \ldots \int 1(|\mathcal{G}(x)|_d \in B, \, \mathcal{N}(0|xU\{0\}) = x) \exp(-\lambda |\mathcal{G}(x)|_d) dx_1 \ldots dx_n .$$

$$(4.4.5)$$

From this one obtains that the conditional distribution of $\mathcal{G}$ given that $\mathcal{C}_d$ has n $(d-1)$-faces is $\Gamma(n,\lambda)$, the Gamma distribution with parameters n and λ, cf. Miles and Maillardet (1982); see also Zuyev (1992).

One proof for this result is based on invariant measure theory (see e.g. Nachbin, 1965; readers unfamiliar with this theory may skip the following details): The measure

$$|\mathcal{G}(x)|_d^{-n} dx_1 \ldots dx_n$$

is seen to be invariant under scale transformations $x \to cx$ in $(\mathbb{R}^d)^n$ with $c > 0$. Hence this measure lifted by the mapping

$$x \to (s(x), w(x)) = (|\mathcal{G}(x)|_d , \, x/|\mathcal{G}(x)|_d)$$

becomes a product measure on the form

$$s^{-1} ds \, \upsilon(dw)$$

because $w(x) = w(cx)$ is invariant and $s(x) = c^{-d} s(cx)$ is equivariant under scale transformations. Consequently,

$$|\mathcal{G}(x)|_d^{1-n} dx_1 \ldots dx_n = ds \, \upsilon(dw)$$

which inserted into (4.4.5) proves the assertion. □

Remark 4.4.2. The distribution of lower-dimensional typical Voronoi facets can at least in principle be obtained in a similar way as for $\mathscr{C}_d$ above. The simplest case is Voronoi edges, as an edge is defined by two Voronoi vertices which in turn are defined by $d+2$ nuclei $x_0, \ldots, x_d$ and $x_1, \ldots, x_{d+1}$, respectively. Note that $x_0 \notin B_0 \equiv B(x_1, \ldots, x_{d+1})$ and $x_{d+1} \notin B_{d+1} \equiv B(x_0, \ldots, x_d)$, see Figure 4.2.2. The remaining nuclei are situated outside $B_0 \cup B_{d+1}$, so the relevant void-probability becomes $\exp(-\lambda |B_0 \cup B_{d+1}|_d)$. Clearly, for any dimension d, we can then reduce the problem of determining the distribution of the length of the typical Voronoi edge to a problem of evaluating a two-fold integral over the angles (θ_0, θ_{d+1}) with $0 < \theta_0 < \theta_0 + \theta_{d+1} < \pi$ as indicated in Figure 4.2.2. Of course this integral depends in a complicated manner not only on $|B_0 \cup B_{d+1}|_d$, but also on the $(d-1)$-content of the 'surfaces' $(\partial B_0) \backslash B_{d+1}$ and $(\partial B_{d+1}) \backslash B_0$, where ∂ denotes topological boundary. For $d = 2$ and $d = 3$, Brakke (1987a, 1987b) has given the details and obtained the results of nu-merical integration as shown in Figures 4.4.3 and 4.4.4.

□

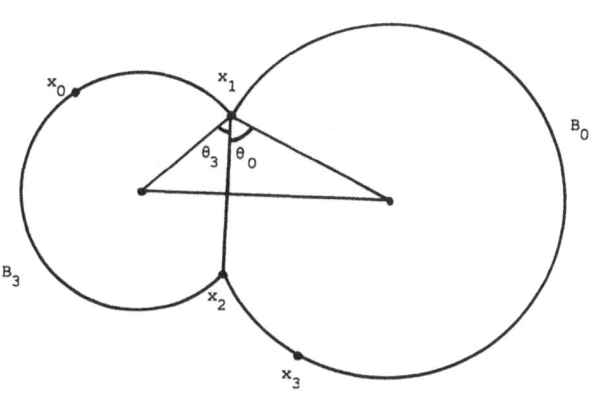

Figure 4.4.2. The case $d = 2$.

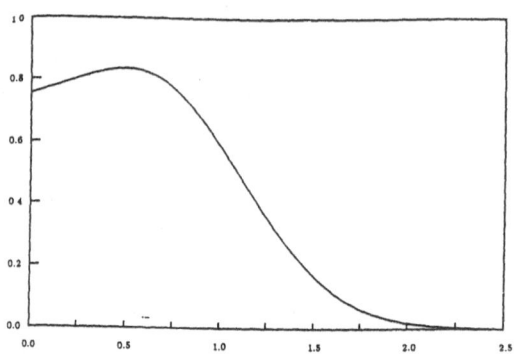

<u>Figure 4.4.3</u>. Probability density function for the length of
the typical Poisson-Voronoi edge when   d = 2   and   λ = 1.
(Reproduced with permission from Brakke (1987a)).

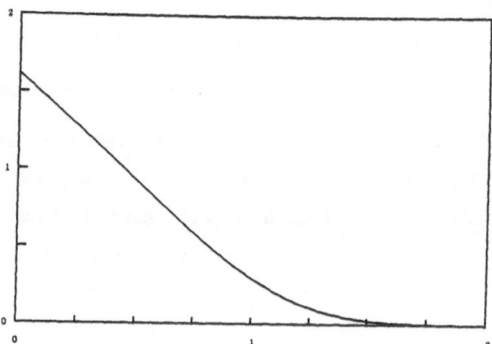

<u>Figure 4.4.4</u>. Probability density function for the length of
the typical Poisson-Voronoi edge when   d = 3   and   λ = 1.
(Reproduced with permission from Brakke (1987b)).

4.5. Simulation procedures for Poisson-Voronoi tessellations and other related models

In this section we present an algorithm due to Quine and Watson (1984) for simulating a homogeneous Poisson process. The points are radially generated, so the algorithm provides an efficient procedure for simulating the typical Voronoi cell as described by Proposition 4.4.1. The algorithm is also tailor-made for simulating a Voronoi tessellation within a ball avoiding the problem with edge effects, when the nuclei are either generated by a homogeneous Poisson process or a related model obtained by e.g. thinning or clustering as described later on in this section. Some emperical results are presented too; other simulated results can be seen in Hinde and Miles (1980), Quine and Watson (1984), Hermann et al. (1989), Møller et al. (1989), and Lorz (1990).

The radial generation of nuclei $\Phi = \{x_1, x_2, \ldots\}$ is determined by independent and uniformly distributed unit vectors $u_1, u_2, \ldots$ on S^{d-1} and independent and uniformly distributed numbers $v_1, v_2, \ldots$ on $(0,1)$ where $\{u_i\}$ and $\{v_i\}$ are independent. Letting $\lambda > 0$ and setting $x_i = r_i u_i$ with

$$r_i = (q_i / (\omega_d \lambda))^{1/d}, \qquad i = 1, 2, \ldots,$$

where

$$q_1 = -\ln v_1, \quad q_{i+1} = q_i - \ln v_{i+1}, \qquad i = 1, 2, \ldots,$$

we shall use the following result.

<u>Proposition 4.5.1</u>. Let $\Phi = \{x_1, x_2, \ldots\}$ be generated as described above, then Φ is a homogeneous Poisson process on $\mathbb{R}^d$ with intensity λ and with nuclei which are radially ordered, i.e. $0 < \|x_1\| < \|x_2\| < \ldots$.   □

Proof. Since $q_1, q_2-q_1, q_3-q_2, \ldots$ constitutes a sequence of independent and exponentially distributed variates with mean 1, $0 < r_1^d < r_2^d < \ldots$ constitutes a Poisson process on $\mathbb{R}_+$ with intensity $\lambda\omega_d$. Hence for a ball B centered at the origin 0 and with radius R, $N_B(\Phi)$ is Poisson distributed with mean $\lambda\omega_d R^d = \lambda|B|_d$. Further, given that $N_B(\Phi) = n$, it is easily verified that $\Phi\cap B = \{x_1, \ldots, x_n\}$ forms a binomial process on B. But any $K \in \mathcal{K}_d$ is contained in B for a sufficiently large R, and so

$$P(\Phi\cap K=\emptyset) = \sum_{n=0}^{\infty} P(N_B(\Phi)=n)P(\Phi\cap K=\emptyset|N_B(\Phi)=n)$$

$$= \sum_{n=0}^{\infty} \frac{(\lambda|B|_d)^n}{n!} e^{-\lambda|B|_d} \left[\frac{|B|_d-|K|_d}{|B|_d}\right]^n$$

$$= \exp(-\lambda|K|_d) .$$

This is in agreement with (4.1.1), whereby the assertion is proved. □

Now, let Φ be a radially generated Poisson process as above. Then with probability one the Voronoi cell $C(n) = C(0|\{0,x_1,\ldots,x_n\})$ is bounded for all sufficiently large n. Let l_n denote the distance to the furthest point in $C(n)$, i.e. the furthest vertex of $C(n)$ provided this cell is bounded. Keeping in mind that Voronoi tessellations result when nuclei start to grow as spheres at the same instant (see Section 1.3),

$$\|x_{n+1}\| > 2l_n \Rightarrow C(0|\Phi\cup\{0\}) = C(n) . \tag{4.5.1}$$

Combining this with Proposition 4.1.1 we obtain the procedure proposed by Quine and Watson (1984): Generate $x_1, x_2, \ldots, x_{n+1}$ until $C(n)$ is bounded and $\|x_{n+1}\| > 2l_n$, and then construct the typical Poisson-Voronoi cell which is given by $\mathscr{C}_d = C(n)$. According to Proposition 3.2.2 this provides also a simulation procedure for the distribution of any translation invariant statistic of the typical k-facet of the Poisson-Voronoi tessellation, $1 \leq k < d$.

Suppose one is interested in the right tail of the distribution of $|\mathscr{C}_d|_d$ or the joint distribution of this statistic and another translation invariant statistic which is positively correlated with $|\mathscr{C}_d|_d$. Then it is more informative to sample from $C_0(\Phi)$ than $\mathscr{C}_d$ where $C_0(\Phi)$ is the Voronoi cell containing 0, cf. Proposition 3.3.2. For the radially generated Poisson process we have that $C_0(\Phi) = C(x_1|\Phi)$, so the Quine and Watson procedure is just modified as follows: Generate $x_1, x_2, \ldots, x_{n+1}$ until $C_1(n) = C(x_1|\{x_1, \ldots, x_n\})$ is bounded and the distance from x_1 to the furthest vertex in $C_1(n)$ is less than $(\|x_{n+1}\|/2) - \|x_1\|$, because then $C_0(\Phi) = C_1(n)$.

Similarly, the Poisson-Voronoi cells with nuclei within a distance R from 0 can be generated. Then we generate $x_1, \ldots, x_{m+1}$ until $\|x_{m+1}\| > R$ and if necessary we generate extra nuclei until all cells $C(x_i|\Phi)$, $i = 1, \ldots, m$, are completed. The Voronoi cells can be constructed as described in Petersen (1987), see also Bowyer (1981) and Watson (1981).

Furthermore, the procedure can also be extended to Voronoi tessellations generated by thinning a radially generated Poisson process $\Phi_\rho = \{y_i\}$ with intensity ρ provided the thinning mechanism of y_i depends only on those points y_j within a fixed distance $\delta > 0$ from y_i. Examples include Matern's first and second type of a hard core process, see

Møller et al. (1989). Let us consider Matern's second type which is obtained by setting

$$\Phi = \{x_i\} = \{y_i \,|\, \|y_i - y_j\| < \delta \Rightarrow a_i < a_j \text{ for all } j \neq i\} \quad (4.5.2)$$

where the a_i are independent and uniformly distributed numbers on $(0,1)$ which are independent of Φ_ρ. Then for each radially generated y_i we need only to simulate extra points y_j, $j > i$, with $\|y_i - y_j\| < \delta$ in order to see if y_i is retained or deleted. Consequently, any Voronoi cell $C(x_i|\Phi)$ is given by $C_i(n)$ whenever $C_i(n) \equiv C(x_i|\{x_1,\ldots,x_n\})$ is bounded and the distance to the furthest vertex of $C_i(n)$ is less than $\|y_{j_n+1}\|/2 - \|x_i\| - \delta$ where j_n is defined by $x_n = y_{j_n}$. This is a convenient stop criteria for simulating the Voronoi cells $C(x_i|\Phi)$ with $\|x_i\| < R$.

Intuitively the Voronoi cells generated by the Matern process Φ in (4.5.2) become more regular than Poisson-Voronoi cells as each cell $C(x_i|\Phi)$ contains the ball centered at x_i and with diameter δ. Note also that Φ is a stationary point process with intensity

$$\lambda = [1 - \exp(-\omega_d \delta^d \rho)] / (\omega_d \delta^d) , \quad (4.5.3)$$

cf. Exercise 4.5.1. A natural parametrization of the class of Matern II hard core processes is provided by $(\lambda,\tau) \in (0,\infty) \times (0,\tau_{max})$, where

$$\tau = \lambda \omega_d (\delta/2)^d = [1 - \exp(-\omega_d \delta^d \rho)] / 2^d \quad (4.5.4)$$

is the packing density, that is the expected proportion of $\mathbb{R}^d$ covered by the nonoverlapping balls centered at the nuclei $\{x_i\}$ and with constant diameter δ. Here $\tau_{max} = (0.5)^d$.

From (4.5.3) and (4.5.4) we obtain the intensity ρ of the Poisson process and the hard core δ used in (4.5.2):

$$\rho = -\ln(1-\tau 2^d)/(\omega_d \delta^d) \quad \text{and} \quad \delta = 2\left[\frac{\tau}{\lambda\omega_d}\right]^{1/d} .$$

The simulation procedure extends also easily to certain types of cluster point process. Let us again consider a specific example, namely a Matern cluster processes. This is obtained by setting

$$\Phi = \bigcup_{y_i \in \Phi_\rho} \{x_{i1}, \ldots, x_{in_i}\} \tag{4.5.5}$$

where each n_i is Poisson distributed with mean μ, $x_{ij} = z_{ij} + y_i$ is uniformly distributed within a fixed distance r from y_i, and Φ_ρ, $n_1, n_2, \ldots, z_{11}, \ldots, z_{1n_1}, z_{21}, \ldots, z_{2n_2}, \ldots$ are mutually independent. Intuitively, this process generates more irregular Voronoi cells than the Poisson process. Further, it is essential that the "offspring" x_{ij} is within a fixed distance from the "parent" point y_i in order to complete a Voronoi cell $C(x_{ij}|\Phi)$ by a finite number of nuclei:

$$C(x_{ij}|\Phi) = C_{ij}(k)$$

when $C_{ij}(k) \equiv C(x_{ij}|\{x_{11}, \ldots, x_{1n_1}, \ldots, x_{k1}, \ldots, x_{kn_k}\})$

is bounded and the distance to the furthest vertex of $C_{ij}(k)$ is less than $(\|y_{k+1}\|/2) - \|x_{ij}\| - r$. Note also that Φ is clearly stationary with intensity

$$\lambda = \rho\mu , \tag{4.5.6}$$

see Exercise 4.5.1, and the class of Matern cluster processes is parametrized by $(\lambda,\mu,r) \in (0,\infty)^3$, where the intensity ρ of the underlying Poisson process is obtained from (4.5.6). Especially, if $r \to \infty$, then Φ tends to a Poisson process with intensity λ.

Finally, let us present some emperical results for simulated 3-dimensional Voronoi tessellations generated by either the homogeneous Poisson process, the Matern II hard core process, or the Matern cluster process. For comparison $\lambda = 1$ in all cases, i.e. the mean volume of the typical Voronoi cell is 1. For the hard core process we used $\tau = 0.12$ which is very close to the maximal packing density $\tau_{max} = 0.125$. For the cluster process we used $\mu = 1/\rho = 5$ and $r = 0.45$. In all the simulations we generated only those Voronoi cells with nuclei within a distance $R = (750/\pi)^{1/3}$ from the origin, i.e. we have on the average 1000 Voronoi cells in a test-ball centered at the origin and with radius R. We have simulated each of the three models in 10 test-balls yielding approximately 10.000 Voronoi cells. Figure 4.5.1 shows the volume distribution of the Voronoi cells for each model: The distribution for the hard core model is nearly symmetric about the mode, while the distributions for the Poisson and especially the cluster model are clearly asymetric with the mode shifted to lower values compared to the hard core model. Also, the variance of the volume distribution is highest in the custer case and lowest in the hard core case. It is evident from Figure 4.5.1 that compared to the Poisson model, the cluster model produces much more irregular cells and the hard core model produces more regular cells. This is less obviously seen in Figure 4.5.2 where central planar sections from 3 test-balls are shown; see also the discussion in Lorz (1990).

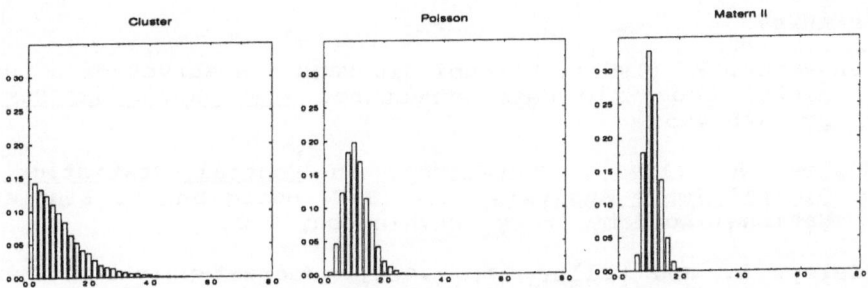

<u>Figure 4.5.1</u>. Histograms for the volume distribution of the Voronoi cells under the Matern cluster, Poisson, and Matern II hard core models, respectively.

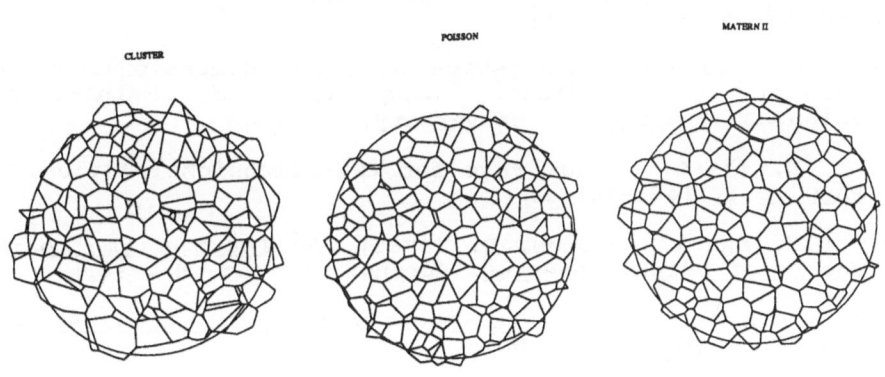

<u>Figure 4.5.2</u>. Central sections through simulated Voronoi tessellations under the Matern cluster, Poisson, and Matern II hard core models, respectively.

<u>Exercise 4.5.1</u>. Verify formulae (4.5.3) and (4.5.6).

References

Aurenhammer, F. (1991). Voronoi diagrams - A survey of a fundamental geometric data structure. Acm. Computing Surveys 23, 345-405.

Baddeley, A. (1991). Stereology. In Spatial Statistics and Digital Image Analysis, eds. J.E. Besag and B. Fingleton. National Academy Press, Washington, D.C.

Baddeley, A. and Møller, J. (1989). Nearest-neighbour Markov point processes and random sets. Internat. Statist. Rev. 57, 89-121.

Boots, B.N. (1987). Voronoi (Thiessen) Polygons. Geo Books, Norwich.

Bowyer, A. (1981): Computing Dirichlet tessellations. The Computer Journal 24, 162-166.

Brakke, K.A. (1987a). Statistics of random plane Voronoi tessellations. Department of Mathematical Sciences, Susquehanna University.

Brakke, K.A. (1987b). Statistics of three dimensional random Voronoi tessellations. Department of Mathematical Sciences, Susquehanna University.

Collins, R. (1968). A sum rule for two-dimensional fluids. J. Phys. C 1, 1461-1472.

Daley, D.J. and Vere-Jones, D. (1988). An Introduction to the Theory of Point Processes. Springer-Verlag, New York.

Delaunay, B.N. (1934). Sur la sphère vide. Bull. Acad. Science USSR VII: Class. Sci. Math., 793-800.

Dirichlet, G.L. (1850). Über die Reduktion der positiven quadratischen Formen mit drei unbestimmten ganzen Zahlen. J. Reine u. Angew. Math. 40, 209-227.

Gilbert, E.N. (1962). Random subdivisions of space into crystals. Ann. Math. Statist. 33, 958-972.

Grünbaum, B. (1967). Convex Polytopes. Wiley, New York.

Hadwiger, H. (1957). Vorlesungen über Inhalt, Oberfläche und Isoperimetrie. Springer-Verlag, Berlin.

Hermann, H., Wendrock, H., and Stoyan, D. (1989). Cell-area distributions of planar Voronoi mosaics. Metallography 23, 189-200.

Hinde, A.L. and Miles, R.E. (1980). Monte Carlo estimates of the random polygons of the Voronoi tessellation with respect to a Poisson process. J. Statist. Comput. Simul. 10, 205-223.

Jagers, P. (1973). On Palm probabilities. Z. Wahrscheinlich keitsth. verw. Geb. 26, 17-32.

Johnson, W.A. and Mehl, R.F. (1939). Reaction kinetics in processes of nucleation and growth. Trans. Am. Inst. Min. Engrs. 135, 416-458.

Kendall, D.G. (1989). A survey of the statistical theory of shape. Statistical Science 4, 87-120.

Lorz, U. (1990). Cell-area distributions of planar sections of spatial Voronoi mosaics. Materials Characterization 25 297-311.

Mecke, J., Schneider, R.G., Stoyan, D., and Weil, R.R. (1990). Stochastische Geometrie. Birkhäuser, Basel.

Meijering, J.L. (1953). Interface area, edge length, and number of vertices in crystal aggregates with random nucleation. Philips Res. Rep. 8, 270-290.

Miles, R.E. (1970). On the homogeneous planar Poisson point process. Math. Biosci. 6, 85-127.

Miles, R.E. (1971). Isotropic random simplices. Adv. Appl. Prob. 3, 353-382.

Miles, R.E. (1974). A synopsis of 'Poisson flats in Euclidean spaces'. In Stochastic Geometry, ed. E.F. Harding and D.G. Kendall. Wiley, New York.

Miles, R.E. (1984). Sectional Voronoi tessellations. Rev. Unión Math. Argentina 29, 310-327.

Miles, R.E. and Maillardet, R.J. (1982). The basic structures of Voronoi and generalized Voronoi polygons. In Essays in Statistical Science, J. Gani and E.J. Hannan, eds., Appl. Prob. Trust J. Appl. Prob. 19 A, 97-111.

Møller, J. (1985). A simple derivation of a formula of Blaschke and Petkantschin. Research Report 138, Department of Theoretical Statistics, University of Aarhus.

Møller, J. (1989). Random tessellations in $\mathbb{R}^d$. Adv. Appl. Prob. 21, 37-73.

Møller, J. (1992). Random Johnson-Mehl tessellations. Adv. Appl. Prob. 24, 814-844.

Møller, J., Jensen, E.B., Petersen, J.S., and Gundersen, H.J.G. (1989). Modelling an aggregate of space-filling cells from sectional data. Research Report 182, Department of Theoretical Statistics, University of Aarhus.

Nachbin, L. (1965). The Haar Integral. Von Nostrand, Princeton.

Okabe, A., Boots, B., and Sugihara, K. (1992). Spatial Tessellations. Concepts and Applications of Voronoi Diagrams. Wiley, Chichester.

Petersen, J.S. (1987). Computation of a 3 dimensional Voronoi tessellation. Department of Theoretical Statistics, University of Aarhus. Unpublished.

Quine, M.P. and Watson, D.F. (1984). Radial simulation of n-dimensional Poisson processes. J. Appl. Prob. 21, 548-557.

Rathie, P.N. (1992). On the volume distribution of the typical Poisson-Delaunay cell. Adv. Appl. Prob. 29, 740-744.

Ripley, B.D. (1981). Spatial Statistics. Wiley, New York.

Rogers, C.A. (1964). Packing and Covering. Cambridge University Press, London.

Santaló, L.A. (1976). Integral Geometry and Geometric Probability. Addison-Wesley, Rading, Massachusetts.

Sibson, R. (1978). Locally equiangular triangulations. Computer Journal 21, 243-245.

Sibson, R. (1980a). The Dirichlet tessellation as an aid in data analysis. Scand. J. Statist. 7, 14-20.

Sibson, R. (1980b). A vector identity for the Dirichlet tessellation. Mathematical Proceedings of the Cambridge Philosophical Society 87, 151-155.

Stoyan, D. and Hermann, H. (1986). Some methods for statistical analysis of planar random tessellations. Statistics 17, 407-420.

Stoyan, D., Kendall, W.S. and Mecke, J. (1987). Stochastic Geometry and Its Applications. Akademie-Verlag, Berlin.

Stoyan, D. and Stoyan, H. (1990). Exploratory data analysis for planar tessellations: Structural analysis and point process methods. Appl. Stoch. Models and Data Anaysis 6, 13-25.

Thiessen, A.H. (1911). Precipitation average for large area. Monthly Weather Rev. 39, 1082-1084.

Van de Weygaert, R. (1991). Voids and the Geometry of Large Scale Structure. Ph.D. thesis, Sterrewacht Leiden.

Voronoi, G. (1908). Nouvelles applications des paramètres continus à la théorie des formes quadratiques. <u>J. Reine Angew. Math.</u> <u>134</u>, 198-287.

Watson, D.F. (1981). Computing the n-dimensinal Delaunay tessellation with application to Voronoi polytopes. <u>The Computer Journal</u> <u>24</u>, 167-172.

Wigner, E. and Seitz, F. (1933). On the constitution of metallic sodium. <u>Phys. Rev.</u> <u>43</u>, 804-810.

Zähle, M. (1988). Random cell complexes and generalized sets. <u>Ann. Prob.</u> <u>16</u>, 1742-1766.

Zuyev, S.A. (1992). Estimates of the Voronoi polygon's geometric characteristics. <u>Random Structures and Algorithms</u> <u>3</u>, 149-162.

Subject and author index

Notation index

General Remarks

Lecture Notes are printed by photo-offset from the master-copy delivered in camera-ready form by the authors of monographs, resp. editors of proceedings volumes. For this purpose Springer-Verlag provides technical instructions for the preparation of manuscripts. Volume editors are requested to distribute these to all contributing authors of proceedings volumes. Some homogeneity in the presentation of the contributions in a multi-author volume is desirable.

Careful preparation of manuscripts will help keep production time short and ensure a satisfactory appearance of the finished book. The actual production of a Lecture Notes volume normally takes approximately 8 weeks.

For monograph manuscripts typed or typeset according to our instructions, Springer-Verlag can, if necessary, contribute towards the preparation costs at a fixed rate.

Authors of monographs receive 50 free copies of their book. Editors of proceedings volumes similarly receive 50 copies of the book and are responsible for redistributing these to authors etc. at their discretion. No reprints of individual contributions can be supplied. No royalty is paid on Lecture Notes volumes.

Volume authors and editors are entitled to purchase further copies of their book for their personal use at a discount of 33.3% and other Springer mathematics books at a discount of 20% directly from Springer-Verlag. Authors contributing to proceedings volumes may purchase the volume in which their article appears at a discount of 20 %.

Springer-Verlag secures the copyright for each volume.

Series Editors:

Professor S. Fienberg
Department of Statistics
Carnegie Mellon University
Pittsburgh, Pennsylvania 15213
USA

Professor J. Gani
Department of Statistics IAS
Australian National University
GPO Box 4
Canberra ACT 2601
Australia

Professor K. Krickeberg
3 Rue de L'Estrapade
75005 Paris
France

Professor I. Olkin
Department of Statistics
Stanford University
Stanford, California 94305
USA

Professor N. Wermuth
Department of Psychology
Johannes Gutenberg University
Postfach 3980
D-6500 Mainz
Germany